A DIVINE LANGUAGE

A DIVINE LANGUAGE

{ Learning Algebra, Geometry, and
Calculus at the Edge of Old Age }

ALEC WILKINSON

FARRAR, STRAUS AND GIROUX
New York

Farrar, Straus and Giroux
120 Broadway, New York 10271

Library of Congress Cataloging-in-Publication Data
Names: Wilkinson, Alec, 1952– author.
Title: A divine language : learning algebra, geometry, and calculus at the
    edge of old age / Alec Wilkinson.
Description: First edition. | New York : Farrar, Straus and Giroux, 2022. |
    Includes bibliographical references.
Identifiers: LCCN 2021062245 | ISBN 9781250168573 (hardcover)
Subjects: LCSH: Wilkinson, Alec, 1952– | Mathematics—Study and
    teaching. | Math anxiety. | Adult learning. | Older people—Education. |
    Learning, Psychology of.
Classification: LCC QA11.2 .W55 2022 | DDC 510.71/5—dc23/eng20220314
LC record available at https://lccn.loc.gov/2021062245

Our books may be purchased in bulk for promotional, educational, or
business use. Please contact your local bookseller or the Macmillan
Corporate and Premium Sales Department at 1-800-221-7945, extension
5442, or by email at MacmillanSpecialMarkets@macmillan.com.

www.fsgbooks.com
www.twitter.com/fsgbooks • www.facebook.com/fsgbooks

1   3   5   7   9   10   8   6   4   2

For

James Wilkinson,

Sara Barrett,

and

Sam Wilkinson

As I made my way home, I thought Jem and I would get grown but there wasn't much else left for us to learn, except possibly algebra.

—HARPER LEE, *To Kill a Mockingbird*

[CONTENTS]

A DIVINE LANGUAGE

Fall

1.

I don't see how it can harm me now to reveal that I only passed math in high school because I cheated. I could add and subtract and multiply and divide, but I entered the wilderness when words became equations and x's and y's. On test days I sat beside smart boys and girls whose handwriting I could read and divided my attention between his or her desk and the teacher's eyes. To pass Algebra II I copied a term paper and nearly got caught. By then I was going to a boys' school, and it gives me pause to think that I might have been kicked out and had to begin a different life, knowing different people, having different experiences, and eventually erasing the person I am now. When I read *Memories, Dreams, Reflections*, I felt a kinship with Carl Jung, who described math class as "sheer terror and torture," since he was "amathematikos," which means something like nonmathematical.

I am by nature a self-improver. I have read Gibbon, I have read Proust. I read the Old and New Testaments and most of Shakespeare. I studied French. I have meditated. I jogged. I learned to draw, using the right side of my brain. A few years ago, I decided to see if I could learn simple math, adolescent math, what the eighteenth century called

pure mathematics—algebra, geometry, and calculus. I
didn't understand why it had been so hard. Had I just
fallen behind and never caught up? Was I not smart
enough? Was I somehow unfitted to learn a logical, complex, and systematized discipline? Or was the capacity to
learn math like any other attribute, talent for music, say?
Instead of tone deaf, was I math deaf? And if I wasn't and
could correct this deficiency, what might I be capable of
that I hadn't been capable of before? I pictured mathematics as a landscape and myself as if contemplating a journey from which I might return like Marco Polo, having
seen strange sights and with undreamt-of memories.

We reflect our limitations as much as our strengths. I
meant to submit to a discipline that would require me to
think in a way that I had never felt capable of and wanted
to be. I took heart from a letter that the French philosopher Simone Weil wrote to a pupil in 1934. One ought to
try to learn complicated things by finding their relations
in "commonest knowledge," Weil writes. "It is for this
reason that you ought to study, and mathematics above
all. Indeed, unless one has exercised one's mind seriously
at the gymnasium of mathematics one is incapable of precise thought, which amounts to saying that one is good
for nothing. Don't tell me you lack this gift; that is no obstacle, and I would almost say that it is an advantage."

I could have taken a class, but I had already failed math
in a class. Also, I didn't want to be subject to the anxiety
of keeping up with a class or slowing one down because
I had my hand in the air all the time. I didn't want a class
for older people, because I didn't want to be talked down
to and more cheerfully than in usual life, the way nurses
and flight attendants talk to you. I could have sat in a class
of low achievers, a remedial class, but they aren't easy to

find. I arranged to occupy a chair one afternoon in an algebra class at my old school, where twelve-year-olds ran rings around me. The teacher assigned problems in groups of five and by the time I had finished the first problem they had finished all of them correctly. They were polite about it, and winning in the pleasure they took in competing with one another, but it was startling to note how much faster they moved than I did. I felt as if we were two different species.

Having skipped me, the talent for math concentrated extravagantly in one of my nieces, Amie Wilkinson, a professor at the University of Chicago, and I figured she could teach me. There were additional reasons that I wanted to learn. The challenge, of course, especially in light of the collapsing horizon, since I was sixty-five when I started. Also, I wanted especially to study calculus because I never had. I didn't even know what it was—I quit math after feeling that with Algebra II I had pressed my luck as far as I dared. Moreover, I wanted to study calculus because Amie told me that when she was a girl William Maxwell had asked her what she was studying, and when she said calculus he said, "I loved calculus." Maxwell would have been about the age I am now. He would have recently retired after forty years as an editor of fiction at The New Yorker, where he had handled such writers as Vladimir Nabokov, Eudora Welty, John Cheever, John Updike, Shirley Hazzard, and J. D. Salinger. When Salinger finished Catcher in the Rye, he drove to the Maxwells' country house and read it to them on their porch. I grew up in a house on the same country road that Maxwell and his wife, Emily, lived on, and Maxwell was my father's closest friend. In the late 1970s, as a favor to my father, Maxwell agreed to read something I was writing, a book

about my having been for a year a policeman in Wellfleet, Massachusetts, on Cape Cod, and this exchange
turned into an apprenticeship. Maxwell was also a writer.
Around the time he spoke to Amie, he was writing *So
Long, See You Tomorrow*, which is the book I give to people
who don't know his work, because it is regarded as one of
the great short novels of the American twentieth century,
and I know that if they like it they will probably like the
rest of his writing. I loved him, and I wanted to know
what he had seen in calculus to delight him. He died, at
ninety-one, in 2000, so I couldn't ask him. I would have to
look for it myself.

The following account and its many digressions is
about what happens when an untrained mind tries to train
itself, perhaps belatedly. It is the description of a late-stage
willful change, within the context of an extended and disciplined engagement, not a hobby engagement. For more
than a year I spent my days studying things that children
study. I was returning to childhood not to recover something, but to try to do things differently from the way I
had done them, to try to do better and see where that led.
When I would hit the shoals, I would hear a voice saying,
"There is no point to this. You failed the first time, and
you will fail this time, too. Trust me. I know you."

After a time my studies began to occupy two channels.
One channel involved trying to learn algebra, geometry,
and calculus, and the other channel involved the things
they introduced me to and led me to think about. While
it was humbling to be made aware that what I know is
nothing compared with what I don't know, this was also
enlivening for me. I am done doing mathematics, so far
as I was able to, but the thinking about it and the questions it raises is ongoing. The structure of my narrative

reflects this dual engagement. It is organized more or less according to the order in which I learned new things, much as in a travel book one visits places along the writer's path.

What did I learn? Among other things, that while mathematics is the most explicit artifact that civilization has produced, it has also provoked many speculations that do not appear capable of being settled. Even those figures occupying the most exalted positions in regard to these speculations can't settle them. A lifetime doesn't seem sufficient to the task.

Some things I had to learn were so challenging for me that I felt lost, bewildered, and stupid. I couldn't walk away from these feelings, because they walked with me in the guise of a gloomy companion, an apparition I could shake only by working harder and even then often only temporarily. There were times when I felt I had declared an ambition I wasn't equipped to achieve, but I kept going.

Finally and furthermore and likewise and not least, I had it in for mathematics, for what I recalled of its self-satisfaction, its smugness, and its imperiousness. It had abused me, and I felt aggrieved. I was returning, with a half century's wisdom, to knock the smile off math's face.

"HOW DO YOU think this will go?" I asked Amie.

"If I had to guess, I would say you will probably overthink."

"How so?"

"X is a useful thing. I can solve for it—I can manipulate it—and I can hear you say, 'What does that *mean*?'"

Do I whine like that, I thought, then I said, "What *does* it mean?"

"It's a symbol that stands for what you want it to stand for."

"What if I don't know what I want it to stand for?"

"See, this is what I'm talking about."

"Well, wait, that's—"

"Here is some advice," she said firmly. "I get it that you try to put things into a framework that you can understand. That's fine, but at first, until you become comfortable with the formal manipulation, you have to be like a child."

She must have seen something in my expression, because she added, "To be a good mathematician you have to be very skeptical, so you have the right temperament." Then, "It's possible I can explain algebra and geometry to you in a way that you'll grasp, but we might have trouble with calculus."

2.

So far as I can tell, mathematicians welcome novices but provisionally. They know if an amateur has trespassed the boundaries of his or her understanding, and they are prone to classifying. This tendency is displayed in the essay "Mathematical Creation" by Henri Poincaré, which appears in the issue of the philosophical journal *The Monist* for July of 1910. It begins, "A first fact should surprise us, or rather would surprise us if we were not so used to it. How does it happen that there are people who do not understand mathematics?"

To Poincaré, mathematics is a matter of reasoning. People can reason through ordinary circumstances, why can't they reason through slightly demanding chains of mathematical symbols when those chains are only smaller and simpler chains connected to one another?

Because people remember the rules only partly, he says, and what they remember they use wrong. More than rules they only half recollect, they should follow a problem's logic.

This had not been not my experience as a boy or now, either. My experience has been that I *might* understand what rule applies, but I don't necessarily understand how to employ it or why it applies in one case and not in another which seems the same case or very closely related. Or I don't know which in a series of rules to use first and in what order the rest should follow. It has been as if I were trying to read but because of some deficiency or inhibition saw only single words without understanding that they formed sentences.

According to Poincaré, most people have ordinary memories and spans of attention. Such people are "absolutely incapable of understanding higher mathematics." Others have a little of the "delicate feeling" necessary to go with powerful memories and spans of attention, and so can master details and understand principles and sometimes apply them, but these people will never create mathematics. A final group, an elite, has the delicate feeling in various degrees and so can understand mathematics and even if their memories are nothing exceptional can create mathematics to the extent that their intuitions have been developed. I am a hybrid of the first and second class, but mainly of the first class, the ordinary one.

In the essay "A Mathematician's Apology," published in 1940, the British mathematician G. H. Hardy writes, "Most people are so frightened of the name of mathematics that they are ready, quite unaffectedly, to exaggerate their own mathematical stupidity," but mathematicians don't usually like it when people say that they can't do math and especially when they say that they don't see the point of trying to. Mathematicians tend to regard such an attitude as taking pride in being ignorant. I don't think that mathematicians realize, though, that what is opaque to many, maybe most, of us is clear to them, and that they have as if been granted special circumstances.

What went on in the minds of the boys and girls whose work I copied, I had no idea until I came across the following sentences in *The Weil Conjectures*, a kind of memoir and imaginative biography of André and Simone Weil by Karen Olsson, who got her degree in math at Harvard. Obscurely echoing Poincaré, Olsson describes feeling as an adolescent that the principles of algebra and geometry "didn't need to be taken in and memorized, the way you had to take in and retain other things. They were, it seemed, already at hand. As though there were simply some latent machine I could turn on with logic, and then! An entire world I never suspected."

I studied music in college and sometimes when math was trampling on me now I would tell myself that it doesn't make sense that someone can learn a complicated practice of one kind and not another, music or languages or philosophy, say, but not math. If the capacity of our brains were the explanation for abilities with numbers, then it would seem strange as a matter of adaptation that some of us had a region of the brain developed for mathematics and some didn't, since there would be a striking and almost disabling deficiency in terms of numbers among those of us who were without the attribute, and in terms of its evolutionary advantage the attribute would appear to be all but inconsequential and maybe even largely useless.

TO PREPARE FOR our meeting, Amie suggested that I read *Algebra for Dummies*, which I had hardly begun when it was borne in on me that it didn't matter who it was for, it was still algebra. I read with a companion self, a twelve-year-old boy who had no desire to sit in algebra class again. Algebra, he reminded me, if you really thought about it, was impossible. We had already proved that. Why prove it twice?

I was surprised to find it difficult. I assumed that in growing older I had also grown smarter. High school math I expected to be totally within my capabilities. I expected to find myself thinking, How could I have found this so challenging?

As a boy I didn't know how to learn anything. So far as I can tell, learning involves an ability to see things consecutively and according to a set of relations. As a child I felt overwhelmed by all that went on around me. I don't think that I was more sensitive than children usually are, but I grew up in a turbulent household. Rather than organize my thoughts, I looked for reasons to avoid having them. By myself in the woods, turning over stones to find salamanders and catching pond turtles, is how I passed much of my childhood. I knew that it was strange to be so solitary, but I didn't know how to be different. I have my own theories about why this is so, but I don't think they are sufficiently diverting to justify lingering any longer in Confession Gulch. As a boy I turned the pages of my algebra textbook and did what part of my homework I could while I waited for a breakthrough, a grace, an illumination that seemed to have arrived for the bulk of my classmates, and I wondered why it hadn't for me.

Because I had been good at arithmetic, I was placed in the advanced math class, which meant that I would take algebra in eighth grade. A few days before the end of summer vacation, I realized that I had forgot how to divide. I thought that if I told anyone, I would have to go to seventh grade again. At the bus stop, on the first day of school, I challenged another boy to divide two big numbers and closely observed how he did it.

I remembered this when I read a mathematician's remark that algebra is a form of arithmetic. My impression was that algebra was less a subject than a practice into which one was inducted by the algebra priests after a series of mortifications. The letters and equations that the teacher drew on the board did not seem related to the numbers I had handled in other classrooms. For one thing, a problem in arithmetic was vertical, one number beneath another, and a problem in algebra, an equation, was horizontal. I felt as if in a permanent present, unable to see how the past and the future were joined. In *Ulysses* James Joyce writes that the present is the drain that the future goes down on its way to becoming the past.

3.

When I read the observation made in 2007 by the Russian mathematician Yuri Manin that algebra was once connected to language, I see some of why I am again having difficulty. The algebra that the ancients knew in Egypt, India, Babylonia, Greece, and Persia is essentially the algebra that is taught in high school. The antique version is called rhetorical algebra, because the problems were described in prose and not in symbols or letters. The version that followed, called syncopated algebra, used abbreviations for common operations. Letters as symbols arrived in the sixteenth and seventeenth centuries, when mathematics began to figure more in commercial and scientific life, and needed to be easier to use and more accurate. The modern version is called symbolic algebra. The symbols are arbitrary. A simple equation tends to be solved for x but it could be some other letter, too. In *Geometria*, published in 1637, Descartes suggested that the earlier letters of the alphabet represent known quantities and the ones after p represent unknown ones, which remains the general practice in teaching. In higher mathematics, though, the conventions are not universal. Amie's field is dynamical systems, which studies the behaviors of a structure such as a solar system that is constrained by particular rules. "In my work," she wrote me, "certain letters are used to symbolize certain things—p, q points; t time; x a numerical unknown; r a positive number; ε, epsilon a small number; μ, mu a measure; z, w complex numbers; A a matrix; M a manifold; n, m integers; X a set, and so on." I appreciated her adding, "and so on," as if she imagined that I were reading along and thinking, That's probably what I'd do, too.

The movement from rhetorical algebra to symbolic algebra resembles the passage from arithmetic to algebra. If no one tells you that you are leaving one field for another and instead behaves as if the fields are the same, even though they appear to be different, it is easy to become confused, at least it confused me. I think this confusion is peculiar to mathematics, which has a quality of otherness.

It seems to be both literal and abstruse. Extremely complicated chains of symbols can express a single, unambiguous thought, whereas language can be made literal only by reducing it to simplest terms, often prohibitory: "Thou shalt not kill." "No smoking." "Keep Off." Computer translations of literary texts rarely satisfy because too many choices are involved, having to do with not only which words to use but also in what arrangement so as to serve the most explicit meaning, let alone the writer's intentions, let alone art. Mathematics is severe and faultless. It became the language of science because of its precision. The theory of relativity can be written in prose, but $e = mc^2$ is more succinct.

ARITHMETIC BECAME ALGEBRA because the ancients found themselves doing repeated calculations to compute, say, the area of a piece of farmland and, while some of the calculations could be done in one's head, if a person thought carefully enough, it was easier to automate them. Eventually a symbol came to stand for the quantity that one was trying to solve for, which is where my difficulties began. Some people are sufficiently comfortable manipulating numbers that manipulating symbols comes naturally to them. They might find the solving of an algebra problem to be something like a game—apply some rules and receive an answer. I simply found it mystifying.

"If you're the kind of person who has trouble keeping track of things, whose mind likes to wander down certain paths and doesn't want to be put on task, like maybe you are, I think solving an algebra problem can be very difficult," Amie said. "The presence of symbols makes it seem that there are too many possibilities."

I wish someone had said on my first day in algebra class, "To start, all you need to know is that you are answering a problem whose solution, instead of involving a single unknown, as it does in arithmetic, involves a second unknown, which we call x." Or, "Algebra is arithmetic, but you just don't know at the beginning what all the numbers are." The arrival of x inserts an abstraction into what in arithmetic had been a literal exchange. The simplest algebra problems (I know now) are only one degree removed from arithmetic. Instead of $2 + 5 = x$, there is, say, $2x + 5 = 9$, which is easy enough to be solved visually, but algebra insists on procedures—subtract 5 from 9 and divide 4 by 2, still arithmetic, but the mind has to accommodate deferring an answer in order to assess what procedures apply and in what order; dividing both sides of the equation by 2 to begin arrives at the same answer but less straightforwardly.

In the first weeks of algebra class, I felt confused and then I went sort of numb. Adolescents order the world from fragments of information. In its way adolescence is a kind of algebra. Some of the unknowns can be determined, but doing so requires a special aptitude, not to mention a comfort with having things withheld. Furthermore, straightforward, logical thinking is needed, and a willingness to follow rules, which aren't evenly distributed adolescent capabilities.

In *Enlightening Symbols: A Short History of Mathematical Notation and Its Hidden Powers*, by Joseph Mazur, I read the following: "At its surface, algebra seems to be the art of manipulating symbols according to some rules for doing so. But then, the modern student knows that all that has to be done is to translate the problem into symbolic notation, and let the rules of symbolic manipulation take it from there." I can appreciate such remarks now, but if someone had said that to me when I was twelve years old, I wouldn't have known if he was messing with me or not.

I found a description of my circumstances in "Mysticism and Logic," by Bertrand Russell. "In the beginning of algebra, even the most intelligent child finds, as a rule, very great difficulty," Russell writes. "The use of letters is a mystery, which seems to have no purpose except mystification. It is almost impossible, at first, not to think that every letter stands for some particular number, if only the teacher would reveal what number it stands for."

Arithmetic is unequivocal. Algebra is a means for making statements that apply widely.

I SAT IN algebra class afraid that I would be called on and handed in homework that I believed was evidence of my dullness of mind. I can understand Russell now when he writes:

> It is not easy for the lay mind to realise the importance of symbolism in discussing the foundations of mathematics, and the explanation may perhaps seem strangely paradoxical. The fact is that symbolism is useful because it makes things difficult.
>
> But how little, as a rule, is the teacher of algebra able to explain the chasm which divides it from arithmetic, and how little is the learner assisted in his groping efforts at comprehension! Usually the method that has been adopted in arithmetic is continued: rules are set forth, with no adequate explanation of their grounds; the pupil learns to use the rules blindly, and presently, when he is able to

obtain the answer that the teacher desires, he feels
that he has mastered the difficulties of the subject.
But of inner comprehension of the processes em-
ployed he has probably acquired almost nothing.

On my second engagement I tell myself that in a problem there is something that I need to find. Looking for it
requires a detachment I can't always enact. I wrote Amie,
"Today I went through several pages of problems and got
every one of them wrong. Very discouraging."

To make me feel better, she wrote, "I am trying to
imagine a similar task I could set for myself. Something
like learning Japanese?"

I told a friend, Deane Yang, who is a professor of mathematics at NYU, how often I was wrong, and he said,
"Getting things wrong is the trick of our trade."

I still don't know what he meant.

4.

It requires unusual abilities to become a mathematician,
that and years of painful training in which the intellect is
forced to bend upon itself.
—David Berlinski, *A Tour of the Calculus*

Reading *Algebra for Dummies*, I am surprised to find that I
recall almost nothing of algebra. I had got lost so quickly
that very little had made an impression. I can still recite
the "Prologue to the Canterbury Tales" in Middle English,
which I was required to learn as a senior in high school. I
remember, "Kingdom, phylum, class, order, family, genus,

species." And that in 585 BCE, Thales predicted an eclipse
of the sun. With algebra, I come up empty.

When I thought I had read a sufficient amount, I went
to Chicago to see Amie. I sat beside her on a couch in
her living room. I held my pencil and notebook ready.
My manner was like that of the novice on his first day
in the monastery poised to have the head monk reveal
how to find God. She said, "I'm not sure where to start."
I had been expecting her to say something like, "There's
a train in Omaha heading for Dallas and leaving at three
in the afternoon." Instead, we sat silently. A dog barked. I
smiled weakly.

There is a belief among certain academics that a subject is less efficiently learned from an adept than from
someone who is studying it or has just finished studying
it. The adept's long acquaintance makes it difficult for
him or her to see the subject in its simpler terms or to appreciate what it is like to approach the subject as a green-
horn. As I sat uneasily beside Amie, it was borne in on
me that I was asking a mathematician with a trophy case
whose standing is international to teach me math that she
had learned nearly half a century earlier as a precocious
child and hadn't used since. Furthermore, she had for the
most part embraced it intuitively and then layered upon
it many other practices, explorations, and diversions. Her
learning had a kind of family tree of associations, and all
I had was what I had picked up piecemeal in a few weeks
of study.

What I might have said to her of the difficulty I was
having was, "Pretend you were a child receiving this information for the first time. Can you remember how you
heard it so that it was sensible to you?" A further complication developed, which is that what is difficult for me

had not been difficult for her, and I don't think she could see why I had such trouble learning what she had found simple. "How do you think you would have thought about this if you hadn't been able to think of it as you had," is the kind of question I would have had to ask, and being philosophical more than practical, it isn't a discussion that would have solved my difficulties. I might have learned something about her, but not likely anything about math.

In *On Proof and Progress in Mathematics*, William Thurston writes, "The transfer of understanding from one person to another is not automatic. It is hard and tricky." We had been working together in a halting way for several weeks when I realized that I was going to have to learn a lot of this on my own.

5.

As a child, I don't think I grasped the concepts of the general and the specific. It seems simple now. Another algebra day-one statement, ideally: in arithmetic the terms are particular; in algebra we are going to make generalizations, meaning we are going to do math without knowing all the terms. In *An Introduction to Mathematics*, by Alfred North Whitehead, I come across simple information that I might have found helpful.

"The ideas of any and of some are introduced into algebra by the use of letters, instead of the definite numbers of arithmetic. Thus, instead of saying that $2 + 3 = 3 + 2$, in algebra we generalize and say that, if x and y stand for any two numbers, then $x + y = y + x$. Again, in the place of saying that 3 > 2, we generalize and say that if x be any number there exists some number (or numbers) y such that y > x."

Whitehead gives five examples of the fundamental laws of algebra:

$$x + y = y + x,$$

$$(x + y) + z = x + (y + z),$$

$$x \times y = y \times x,$$

$$(x \times y) \times z = x \times (y \times z),$$

$$x \times (y + z) = (x \times y) + (x \times z).$$

The first is the commutative law of addition; the second is the associative law of addition; the third and fourth are the commutative and associative laws of multiplication; and the fifth is the distributive law of addition and multiplication. As for the concision that symbols provide, Whitehead writes that in prose the first rule, instead of being four letters, would be, "If a second number be added to any given number the result is the same as if the first given number had been added to the second number."

I forced myself to advance. I say advance, but sometimes only time was advancing. My progress might be sideways and sometimes backward. On the other hand, I had no standard to compare myself to. I had never known an older person who was trying to learn math. Older self-improvers usually memorize poetry or study a language, which they can practice with other people. I was able only to sit by myself in a room and review mathematical rules and terms in the hope of making them familiar. I won't say that it was like learning prayers, but it had an in-the-service-of feeling, as if I were secluded. It was similar in that prayers, like mathematical procedures and principles, have specific applications.

Meanwhile, a part of me was resisting the effort, perversely, as if there were a pleasure in failing, or at least obstructing, even if a sour one. The energy being claimed by the resistance I might have used for the task, and until I had it, I wouldn't be firing on all cylinders. Other days the problems tipped over like targets in a shooting gallery, and I went ahead intrepidly.

6.

Possibly not everyone knows that algebra is thought to be the contribution, although maybe not entirely the invention, of a Persian mathematician and librarian named Muhammad ibn Musa al-Khwarizmi, who lived in Baghdad in the ninth century. Al-Khwarizmi wrote a book called *Al-Jabr W'al Muqabalah*, which translates to "Calculation by Restoration and Reduction." *Al-Jabr* has been translated to *algebra* and is the first time the word appears. In *Imagining Numbers*, the mathematician Barry Mazur says that *al-jabr* and *al-muqabalah* also refer to processes. "*Al-jabr* is the operation of moving quantities from one side of an equation to the other," he writes, and "*al-muqabalah* is the operation of collecting 'like' terms."

Algebra perhaps had antecedents, according to a scholar named Peter Ramus, writing in the sixteenth century. In the entry for *algebra* in the eleventh edition of the

Encyclopedia Britannica, Ramus is the source for the assertion that "there was a certain learned mathematician who sent his algebra, written in the Syriac language, to Alexander the Great, and he named it *almucabala*, that is, the book of dark or mysterious things," which is a pretty good title for a book about algebra.

For a reason I don't know, perhaps from taking things too literally, historical accounts of algebra often mention that *Jabr* is from the verb *jabara* and means "to join" and that an *algebrista* in Spain was a "bone-setter." Al-Khwarizmi wrote that his book concerned "what is easiest and most useful in Arithmetic, such as men constantly require in cases of inheritance, legacies, partition, lawsuits, and trade, and in all their dealings with one another, or where the measuring of lands, the digging of canals, geometrical computations, and other objects of various sorts and kinds are concerned."

I am not alone in finding algebra largely incomprehensible. Darwin, another self-improver, studied mathematics with a tutor during the summer of 1828, when he was nineteen. "The work was repugnant to me, chiefly from my not being able to see any meaning in the early steps in algebra," he writes in his autobiography.

In the theater of my mind, my adult self was prepared to step in with an I'll-handle-this attitude to defend the boy I had been against mathematics. As an older, somewhat educated person, I could see that mathematics was wrong, because based on illogical and inconsistent propositions. It was taught to children because they are impressionable. Adults would see through it. The rules and procedures pressed on children amounted to an indoctrination and were not rules so much as articles of faith.

I carried this attitude into my renewed engagements.

I practically enameled myself with it. When algebra wouldn't yield, I adopted, as Amie had predicted, a position of over-literalness, which viewed math skeptically, as a con, really. The more overwhelmed I became, the more I insisted that math submit to being interrogated. I believed that I could refuse to accommodate math's self-serving willfulness. Why anyone had tolerated it was a question I couldn't answer. It seemed like being a mathematician was like being in a cult. In exchange for accepting a wagonload of irrational claims, you lived in a perfectly ordered world.

Amie I figured had agreed to these arrangements before she was old enough to see that they were unfounded. Because they were told to her by adults whom she trusted, she accepted them, and had lived for years under principles that she never had realized were unsound. I looked forward to disabusing her, which I think was unworthy of me, but I also felt sympathy for her situation. To have one's lifelong assumptions overthrown in middle age is not a simple matter. It needed to be done with care and consideration. Tread carefully, I thought.

7.

Misreading a symbol or failing to register the meaning of one, I am sometimes lost for days, left in the dust of the algebra train as it heads for the horizon without me. It is a commonplace that to the degree that mathematics is an imaginative pursuit it is also an art, but such a thing does not happen to me in the other arts. I can find pleasures in a book or an artwork or a piece of music that I don't completely understand. In any other serious field of imaginative work there is no necessarily correct interpretation, but in mathematics you must be certain. Eventually, what you don't know will stop you, ask for your papers, and detain you for questioning.

Practicing other arts, one can proceed, at least a certain distance, as an innocent and even blindly. A mathematician can also proceed blindly, but not as a novice. The mists and darknesses are only for adepts. As a writer, a painter, or a musician your limitations assert themselves sooner or later, but you might go a ways before they do. Occasionally, they become part of your style. I read once of David Hockney's answering the question, Why do your shoeless figures always have socks on, by saying, I can't draw feet. You can't do math without an awareness of what is behind you, the stately progressions, the panorama of understandings, findings, and breakthroughs. Mathematics is rigid, but for those who comprehend it, the rigidity becomes liberating, a kind of touchstone from which you can launch journeys and to which you can confidently return. Math is modern and historical at the same time. Nearly all beginner math—that is, algebra, Euclidean geometry, and calculus—was known in the eighteenth century and in the case of algebra and Euclidean geometry is ancient.

As for inconsistencies, I collided early with pi, which has multiple and unconnected uses. In fact it seems nearly ubiquitous, an apparition hovering between the background and foreground in a multiplicity of mathematical statements, which I regard as suspicious. Its ubiquity makes it appear to have no real identity; it seems more like a placeholder than a real thing. Perhaps like me not everyone recalls that pi, an endless number beginning 3.141, is the ratio of a circle's circumference to its diameter.

It is also the equivalent of 180°, because $\pi = c/d$, circumference over diameter. A diameter is equal to twice the radius, the radius being the line from the center of a circle to its edge, so diameter equals 2r. By convention, a circle's circumference is 360 degrees when the radius is equal to 1. Pi then equals 360°/2r or 180°. Pi makes so many other appearances in mathematics, though, and is useful in so many situations that it seems absurd. It seems like a mathematician at a loss just writes something like, "Thus we see that," and adds pi, and it's quitting time.

Amie told me that pi appears so often because many mathematical formulas derive their features from repeating patterns. They exemplify pi, but not in ways that are always immediately apparent.

"Have you ever heard of Buffon's Needle?" she asked. "It's another way of computing pi. You take parallel lines on a piece of paper. You space them an inch apart, because that's exactly the length of your needle. Then you toss the needle into the air like you would a coin, and you count the number of times it hits one of the lines. It will never hit both, because it's exactly the length of their separation. That's a probability zero event. Anyway, you divide the number of hits by the number of tosses. That answer's going to be some proportion less than one, I believe it's 2/pi. There is a lot of symmetry in this problem. Pi has to do with anything that involves periodicity or cyclical behavior."

This only deepened my reservations.

PROBABLY HUNDREDS AND maybe thousands of mathematicians have written books about mathematics. I wasn't equipped to write from a scholarly perspective or an insider's one, either. As an older novice, but one with a certain acuity, I hoped I might see things that mathematicians had overlooked as being too familiar or perhaps had never even noticed.

It wasn't long before I had such an insight. I was staring at a page of equations, and it was borne in on me that mathematics is a language and an equation is a sentence. The subject appears on the left-hand side of the equal sign, along with the verb in the form of the transaction being conducted, and the figure on the right-hand side is the object: two old friends, $4x$ and 5, go walking and meet 5's uncle, 13.

I was pleased with this insight, which struck me as deep and even lyrical. It cheered me in light of the discouragements I had encountered and reinforced my sense of purpose.

8.

Mathematicians know what mathematics is, sort of. I have heard: mathematics is the craft of creating new knowledge from old using deductive logic and abstraction. The theory of formal patterns. Mathematics is the study of quantity. A discipline that includes the natural numbers and plane and solid geometry. The science that draws necessary conclusions. Symbolic logic. The study of structures. The account we give of the timeless architecture of the cosmos. The poetry of logical ideas. Statements related by very strict rules of deduction. A means of seeking a deductive pathway from a set of axioms to a set of propositions or their denials. A science involving things you can't see whose presence is confined to the imagination. A proto-text whose existence is only postulated. A precise conceptual apparatus. The study of ideas that can be handled as if they were real things. The manipulation of the meaningless symbols of a first-order language according to explicit, syntactical rules. A field in which the properties and interactions of idealized objects are examined. The science of skillful operations with concepts and rules invented for the purpose. Conjectures, questions, intelligent guesses, and heuristic arguments about what is probably true. The longest continuous human thought. Laboriously constructed intuition. The thing that all science, as it grows toward perfection, becomes. An ideal reality. A story that has been being written for thousands of years, is always being added to, and might never be finished. The largest coherent artifact that's been built by civilization. Only a formal game. What mathematicians do, the way musicians do music.

Bertrand Russell said that mathematics, by its nature as an explorative art, is "the subject in which we never know what we are talking about, nor whether what we are saying is true." Darwin said, "A mathematician is a blind man in a dark room looking for a black cat which isn't there." In *Alice's Adventures in Wonderland*, Lewis Carroll has the Mock Turtle say that the four operations of arithmetic (addition, subtraction, multiplication, and division) are ambition, distraction, uglification, and derision. A complicating circumstance is that mathematics, especially in its higher ranges, *is* hard to understand. It begins as simple, shared speech (everyone can count) and becomes specialized into dialects so arcane that some of them are spoken by only a few hundred people in the world.

No scripture is as old as mathematics is. All the other sciences are younger, most by thousands of years. More than history, mathematics is the record that humanity is keeping of itself. History is subjective and can be revised or manipulated or erased or lost. Mathematics is objective and permanent. $A^2 + B^2 = C^2$ was true before Pythagoras had his name attached to it, and will be true when the sun goes out and no one is left to think of it. It is true for any alien life that might think of it, and true whether they think of it or not. It cannot be changed. So long as there is a world with a horizontal and a vertical, a sky and a horizon, it is inviolable and as true as anything that can be thought.

Mathematicians live within a world that is invincibly certain. The rest of us, even other scientists, live within one where what represents certainty is So-far-as-we-can-tell, this-result-occurs-almost-all-of-the-time. Because of mathematics' insistence on proof, it can tell us, within the range of what it knows, what happens time after time.

As precise as mathematics is, it is also the most explicit language we have for the description of mysteries. Being the language of physics, it describes actual mysteries—things we can't see clearly in the natural world and suspect are true and later confirm—and imaginary mysteries, things that exist only in the minds of mathematicians. A question is where these abstract mysteries reside, what is their home range. Some people would say that they dwell in the human mind, that only the human mind has the capacity to conceive of what are called mathematical objects, meaning numbers and equations and formulas and so on, the whole glossary and apparatus of mathematics, and to bring these into being, and that such things arrive as they do because of the way that our minds are structured. We are led to examine the world in a way that agrees with the tools that we have for examining it. (We see colors as we do, for example, because of how our brains are structured to receive the reflection of light from surfaces.) This is a minority view, held mainly by neuroscientists and a certain number of mathematicians disinclined toward speculation. The more widely held view is that no one knows where mathematics resides. There is no mathematician who can point somewhere and say, "That is where math comes from," or "Mathematics lives over there," while maybe gesturing toward magnetic north and the Arctic, for me a suitable habitat for such a coldly specifying discipline.

The belief that mathematics exists somewhere else than within us, that it is discovered more than created, is called Platonism, after Plato's belief in a non-spatiotemporal realm that was the region of the perfect forms of which the objects on earth were imperfect reproductions. By definition, the non-spatiotemporal realm is outside time and space. It is not the creation of any deity, it simply is. To say that it is eternal or that it has always existed is to make a temporal remark, which does not apply. It is the timeless nowhere which never has and never will exist anywhere but which nevertheless is. The physical world is temporal and declines, the non-spatiotemporal one is ideal and doesn't.

A third point of view, historically and presently, for a small but not inconsequential number of mathematicians, many of them exalted ones, is that the home of mathematics is in the mind of a higher being and that mathematicians are somehow engaged with Their thoughts. Georg

Cantor, the creator of set theory, which in my childhood was taught as part of the "new math," said, "The highest perfection of God lies in the ability to create an infinite set, and its immense goodness leads Him to create it." And the wildly inventive and self-taught mathematician Srinivasa Ramanujan, about whom the movie *The Man Who Knew Infinity* was made in 2015, said, "An equation for me has no meaning unless it expresses a thought of God."

In Book 7 of *Republic*, Plato has Socrates say that mathematicians are people who dream they are awake. I partly understand this, and I partly don't.

IN THE ISSUE of the *Bulletin of the American Mathematical Society* for October 2007, Terence Tao, whom many of his contemporaries regard as the greatest living mathematician, whatever that means, published a paper called "What Is Good Mathematics?" It might be mathematics that solved a complex problem, Tao says. Or used a technique handsomely or invented a new one. Or combined areas that hadn't been combined before. Or one that by means of insight simplified an important concept or found a unifying one. Or discovered something not known before (and it interests me that he uses discovered instead of created). One with an important application to fields other than math, such as statistics or computer science. One that illuminated an argument. Or was visionary and had the potential to provoke new work. Or was rigorous. Or beautiful, or elegant, or creative, or useful, or strong. Tao regards all of these circumstances as consequential and does not arrange them in a hierarchy.

Fields are not inherently robust, Tao says. He imagines

a field might become "increasingly ornate and baroque." Or have brilliant conjectures but few prospects for proofs. Or consist of a collection of problems without a theme that unifies them. Or one that has lapsed into pure technique and grown lifeless. Or another grown lifeless because it restates only classical truths, even if more simply, elegantly, or succinctly.

WITHIN A MATTER of weeks, I had read that equations were sentences and mathematics was a language so often that I realized they were clichés. This was deflating.

9.

The mathematician Alonzo Church, who taught Alan Turing, told another of his students, David Berlinski, "Any idiot can learn anything in mathematics. It requires only patience." I would sit with a pencil and paper trying to solve an algebra problem and sometimes I could go only so far before my mind would halt, because I had used up what little I knew that might apply. It hadn't occurred to me to think of algebra as the bright boys and girls I had been among had thought of it, as a series of related procedures. They were constructing a map. I was collecting postcards from places where anxiety or incuriousness had kept me from leaving my hotel.

In my second encounter I was subject to the same limitations with respect to equipment and attitude that I had been subject to as a child. No matter how we change or what happens to us, something fundamental seems to insist on staying the same or at least coming along for the ride.

Solving algebra problems, I knew what was to happen— I was to find the number represented by x—but I couldn't without understanding the rules to reveal what x stood for. This is a consequence of there being only one answer to a student math problem, which is designed less to make a person think than to use theorems and properties learned lately. It occurs to me only now that when I was a boy I might have looked at the lesson that the problems were illustrating and understood that the homework was to be solved by those methods. The textbook's structure was deliberate and not arbitrary, as it seemed to be. The problems on page 164 were asking me to use the information from pages 162 and 163, not from page 34 or information I would not receive until page 310. I saw my math textbook less as a practical manual than as an inscrutable text in which the answer I needed might be anywhere and might even be concealed.

Amie's daughter, Beatrice, studying math at Harvard, told me, "Math is about learning patterns, so that you don't have to relearn them. You have to understand that everything in math is connected." The question of whether one sees patterns in a discipline's design has partly to do with whether one's neurology equips one to see them and partly with whether one is made aware from an early age that there are patterns to be seen.

Sometimes now, as Russell described, I had an experience that I didn't have as a boy, that of solving a problem in a procedural fashion by having found the pattern that the problem was asking me to find, even though I didn't understand the reason for there being a pattern or why the pattern applied to the case. This was a form of cheating, of acquiring a serviceable method without a foundation, but it typically followed such periods of frustration that I felt entitled to it.

After a while I began to see that the limitations I thought I was uncovering in mathematics were not flaws, they were examples of the limited range of my thinking, and perhaps of my ability to think at all. I decided I might benefit from being more receptive. One morning I sat trying to believe that the means for solving a problem might appear from anywhere within the circle of my awareness and not only from where I had already insisted that it couldn't, as if I might see it from the corner of my eye, which for a while made doing algebra a little like attending a séance.

I came to realize what perhaps should have been obvious, that I was creating difficulty by behaving perversely. I had been personifying math as a spiteful thing, and as a result it was demonstrating the capacity to behave spitefully. In a kind of half-smart, half-cunning way I believed that math had tripped me up before, and I wasn't going to allow it to again. I was permanently on alert for deceitful propositions. I was so occupied with this ridiculous position that I didn't see that there were more sensible means of engagement.

I am aware that this is not a profound realization, but it was important to me, and I see no reason to pretend to be smarter than I am. It led me to realize that while I had learned a lot of things in my grown-up life, I had spent very little time trying to learn something that was difficult for me to learn.

Studying algebra required remembering the math that I had been taught before algebra, meaning the properties of fractions, negative numbers, exponents, and decimals. If I ever knew, I had forgot how to divide fractions or what happens when a fraction is raised to a power or what it means when there is an exponent in the denominator or what to do when the exponent is negative. So for a while I had to put algebra aside and sit in the remedial room.

To review percentages I was in sixth grade, according to the texts, which, especially as I found percentages difficult, was lowering to my self-esteem. I was trying to find where I had lost my way, and I hoped not to get lost again. I was trying to assemble a set of reliable assumptions with which I could safely advance.

10.

Now I think being wrong almost all the time for weeks is sort of funny, but I didn't then. It hadn't occurred to me that not only might I find algebra difficult, but maybe algebra actually was difficult. Doing algebra again was like meeting someone you hadn't seen in years and being reminded why you really never liked him or her. To learn, I had to begin to take my chances, to think of myself as good at some things, maybe, not good at others, surely, and to apply myself, with a degree of humility, to being edified by something that was more demanding, more severe, deeper, and less tolerant than I had expected. I knew that by accepting my shortcomings I would be able to proceed without pretensions or pressures, although I risked finding out that I had no capacity to learn math and that the shortcoming was proof of my limited intellectual powers. I would rather have had algebra on the ropes.

JUST AS THERE are impairments in reading such as dyslexia, there are impairments in mathematics. The mathematical version of dyslexia is called dyscalculia. Someone with dyscalculia cannot tell from a glance how many objects are in a small group. They also tend to have difficulty reading clocks and sheet music and in estimating how far away an object in the distance is.

I don't have dyscalculia, but doing math as a boy made me anxious, and it still does, a little. Nobody called it math anxiety then, but they do now. It's a syndrome. Math makes some people apprehensive to the point of dread. A severe aversion to mathematics is called high math anxiety, HMA, and also math trauma, which sounds overdramatic to me, but not everyone thinks so. HMA apparently makes a person's heart beat faster. The amygdala is part of the brain's limbic system, where, among other things, emotional responses and instincts for survival reside. From the paper "Avoiding Math on a Rapid Timescale: Emotional Responsivity and Anxious Attention in Math Anxiety," by Rachel Pizzie and David Kraemer, I learned that someone with high math anxiety responds to math with "aversive, distancing behavior and increased threat-related amygdala reactivity," so it wasn't all in my head, although it was.

HMA, it turns out, is widespread. It afflicts about 10 to 15 percent of college students, and about 20 percent of the rest of us. (By comparison, social anxiety afflicts about 6.8 percent of us, and 3.1 percent of us have generalized anxiety disorder.) Grade school boys tend to have more math anxiety than grade school girls, but in college, women tend to have slightly more than men, although this may have to do with the obstacles that woman math students have traditionally faced from male professors. Until 1992, when a sufficient number of people complained, Barbie dolls said, "Math class is tough," but Ken dolls didn't.

You can have math anxiety because you don't do well at math, but you can also not do well at math because you have math anxiety. This circumstance is called bidirectional. Moreover, math anxiety is contagious. In India, when parents with high math anxiety helped their children do math homework, some of the children got math anxiety. Through gestures and attitudes conveying unease, a teacher with math anxiety can afflict students.

A test called the Abbreviated Math Anxiety Scale identifies HMA. A person ranks experiences such as watching a math teacher diagram a problem on a blackboard, starting a new chapter in a math textbook, and being given a math quiz unexpectedly. There are also statements such as "Working on mathematics homework is stressful for me," and "I get nervous when taking a mathematics test," and "I believe I can think like a mathematician," to which one answers never, seldom, sometimes, often, or usually. Lesser diagnoses of high math anxiety include low, some, moderate, or quite a bit.

I answered a sufficient number of questions on an HMA test I found online to qualify as having some math anxiety, but I don't know how much, because I didn't take the whole test. I am comfortable regarding myself as math averse or maybe math resistant.

Anxiety in relation to curriculum and subject isn't common. With the exception of physics, in which mathematics figures, there isn't science anxiety generally, at least I have never heard anyone say so, and there isn't literature anxiety or history anxiety, either. Remembering that the Battle of Hastings was fought in 1066 or that *Crime and Punishment* is about a murderer is a simple task of recall. Doing mathematics is a serial task, involving several components of memory and various rather than single areas of the brain. Reading "Math Anxiety and Its Cognitive Consequences," whose lead author is Mark Ashcraft, I find that the first component activated by a math problem is the search for a means of finding an answer. A second component seeks the tools, and a third employs them. The more steps to a mathematics problem, the more someone with HMA might think, Quit.

Someone with HMA does mathematics in a state of vigilance. He or she has an habitual response that occurs so quickly and offhandedly that it hardly registers in consciousness. According to Pizzie and Kraemer, this response is similar to "the response of phobic individuals to phobic stimulus." The abrasion of the engagement interferes with absorbing what is meant to be learned. Since mathematics advances by a progression of methods and just gets harder, math anxiety, like seasickness, doesn't go away, and not infrequently it worsens. For mathematically disabled children, math anxiety may reach a climax around ninth grade, with algebra, to which I can only say, I *knew* it!

I remember feeling nervous in algebra class and then, as I fell behind, embarrassed and determined not to let anyone know. Children compare themselves with other children, and how they feel about themselves affects who they think they can be friends with. Also, what they think they can do, and who they imagine they are.

In "The Classroom Environment and Students' Reports of Avoidance Strategies in Mathematics," the lead author of which is Julianne C. Turner, I find remarks by a University of California professor named Martin Covington saying that for many students, "to be able is to be worthy, but to do poorly is evidence of inability and is reason to despair."

The attempts I made to keep people from knowing that I didn't understand math are common strategies that Covington calls "ruses and artful dodges," the purpose of which is "to escape being labeled as stupid." Adolescence is a period of secret-keeping, and it didn't occur to me to wonder whether anyone else was having as difficult a time as I was. The boys and girls whose ruses had failed obviously, the dunces, I pretended I had nothing in common with.

11.

A grievance now, a pedagogical one, maybe more than one: writers of algebra textbooks appear to take their readers to be more accomplished than I am, and so they skip steps in a collegial kind of you-and-I-know-that-these-steps-are-too-simple-to-include spirit, wink-wink, but it isn't too simple for me, who is following line by line and doesn't expect skips. A failure to follow can cost me hours, because I am dogged and also because I don't always realize that a step is being skipped, I just know that the reasoning I think I am pursuing no longer makes sense. By the time I hear from Amie what the writer has left out I am good and worked up, I am hot to trot.

The omissions make me suspect that a math joke is being played on me, the plodding specifist. I end up with three or four textbooks open and writing Amie as if sending dispatches from some beleaguered outpost and needing an answer right away, because everything has broken down. Since I was typically so annoyed and frustrated, my approach usually took the form of an assault, with algebra in the crosshairs—goddam you, I'll *force* you to obey straightforward terms. I was feeling an adolescent's retributive anger, fed by slights and resentments, vehement and close to irrational, nurturing bruised feelings, and instigated by the world's dishonesty. I understand how I had found it easy to walk away from math. I had given it a chance. It was a brute, malign, and mechanical thing, hostile to innocence and hope. I did not start out being mathematically averse, I remind myself. I did well for a time, with arithmetic. I was assaulted and weeded out, and how was that fair?

No one likes getting things wrong in public. It was not fun to be called on and to demonstrate not only that I did not have the right answer, I didn't even know how to arrive at it. On the worst occasions, I didn't even know what question I was being asked.

One day I sat in a class that Amie was teaching to undergraduates on linear algebra, a subject taught after calculus. It took place in a room like a small theater with rows of seats rising toward the back wall. For an hour and a half I had no idea what she was talking about. Occasionally she wrote equations on the blackboard and then she would turn and ask the class, "Do you understand?" I asked her afterward how she could tell if they were being truthful. "I can see it in their faces," she said.

As a boy in math class I would take a seat toward the

back of the room. It hadn't occurred to me that my face might reveal something I was trying to keep secret. I just didn't want to be asked for an answer.

THERE IS A degree of uneasiness in learning a subject that taxes one's capacities. There is additional uneasiness if one lost to it on the first encounter. I worry not only, Am I up to it, but, I wasn't up to it before, will I be up to it now? And how much of having not been up to it was from how I was taught and how much rests with me and my ableness to learn?

I thought I should use the algebra textbook being used at the school I went to. This book has at the beginning the sentence "A variable is a letter used to represent one or more numbers." I would find it easier if the sentence were written, "In algebra, when you don't know a number, you use a letter, called a variable." In the phrasing of the textbook explanation, questions arise for me: How do I use a variable? In what circumstances? Does one letter represent more than one number, or do I need additional letters? In my version I am presented with a rule and am prepared to understand a following rule: when more than one number is unknown, you will use more than one letter, one for each unknown.

These textbooks are enormous. They are written by many people. There is a general boosterish quality to the prose, as if learning math is not only *fun!* but also obscurely patriotic, the duty of an adolescent citizen-in-waiting. I acknowledge that there is a great deal of material to consider in the writing of a textbook. To do it succinctly and well would be an achievement of good thinking and good writing. It seems an irony that the most precise of sciences is often presented imprecisely. Whether the textbooks intend it or not, by being unrigorous, they make math more complicated and obscure than is necessary. A rigorous program would allow a plodder such as me to follow each step and build confidence. When I had trouble with one textbook, I found another, but nearly all of the books were poorly written. In addition to leaving things out, they were careless about language, their sentences were disorderly, their thinking was frequently slipshod, and their tone was often cheerfully and irrationally impatient.

I felt asked by these books to solve puzzles that appeared to depend on a cliquish, arcane knowledge, one that was, in fact, apprehensible if explained in an uncomplicated way, using clear thinking and plain speech. It was partly opaque because their ham-handed writing and slovenly thinking made it so. Furthermore, it seemed to flatter the writers' vanity to regard themselves as keepers of rituals and secrets. Occasionally I could find a principle illustrated simply in another book, and then I could go back and write, "Obscenity you, jerkoff," in the margin of the first book, since I owned it, and in eighth grade it was the property of the school district.

12.

I planned to allow six weeks to learn algebra. Studying six hours a day, six or seven days a week, I would spend about as much time as a student spends in a classroom in a year. The weeks began to pile up, though, and, while I noted the pages left in the textbook, I began to think, How can I have difficulty understanding what twelve-year-olds can? An exchange of emails I had with Amie ("This example is wrong!" "It's not wrong, it's trying to show you a different way of handling the problem") concluded with her writing, "It's been really instructive conversing with you about your readings. It seems maybe that you assign too much authority to the author, that you expect to be told the stone-cold truth and what to do at every step, whereas the authors are dwelling on possible exceptions and uncertainties before making the truth and the main point clear."

If I wasn't to grant authority to the author, though, whom was I to grant it to? On the other hand, I saw, once again, that the more obstinate I grew, the more I missed the point. I was continuing, rather flamboyantly, to enact Amie's belief that I would overthink. Actually, I was brushing up against paranoia. I felt I had to follow each phrase carefully, lest the meaning of a sentence be lost, and I lead myself into overthinking. It was like going into a skirmish with all your soldiers arguing with one another.

I might have felt better if I had shed this stance, or exchanged it for another, but I didn't know what to exchange it for. My sense of myself was caught up in the effort. I was trying to learn something that I worried my intelligence did not equip me for, and I was really, even if I'm a little embarrassed to say it, afraid that I would fail totally unless I defended myself.

I had begun with a sort of math-works-for-*me*-now attitude, and math had answered, *Nuh-uh*. I wasn't going to have to work for algebra, but I was going to have to get along with it. That threw me against all the uncertainties that had so plagued and harassed my adolescent self, and I wasn't eager to subject myself to them again.

At a party I ran into a colleague from *The New Yorker*, Calvin Trillin, who asked me what I was working on. I said a book about mathematics. He looked at me closely and said, "For or against?"

13.

Years ago I listened to a philosopher engage in an argument at a friend's apartment. Instead of defending a position, he cared only to know more and to understand the other person's point of view. The way he was arguing may be common among philosophers, but I hadn't heard anyone take part in an exchange while evincing such humility and receptivity. His example became an aspiration for me, although I have never been able to enact it, too hotheaded and too easily baited. Nevertheless, it was borne in on me that I was going to have to personify some element of it if I was going to learn algebra. Unless I could come up with some proof, I knew I couldn't continue to call Amie and complain about algebra's seeming to be irrational.

One day I called Amie with yet another inquiry and realized that I had framed it not as a demand or an accusation but in such a way that it allowed an explanation. For perhaps the first time, I hadn't insisted that Amie abandon mathematics and confirm my objections, but had allowed myself to think that in order to understand the matter, I would have to follow a line of reasoning different from a defense of my position. I had to disengage myself from believing that my identity was attached to my being right. That I was able to, *barely*, was a surprise to me.

ONE OF THE first procedures to defeat me involved multiplying polynomials, expressions that contain more than two terms. The solution to $(1 + x - y)(12 - zx - y)$, a problem I was given in *Algebra*, by I. M. Gelfand and A. Shen, involves multiplying each term in the first parentheses by each term in the second: 12 times $1 = 12$; 12 times $x = 12x$; 12 times $-y = -12y$; $-zx$ times $1 = -zx$; $-zx$ times $x = -zx^2$; and so on. Then you combine those terms that can be combined. Each problem seemed to rely on a different assumption. I would think that I had learned a rule for combining terms, for example, but when I applied it my answer was wrong. In this case, Amie pointed out that I was trying to combine terms that were incompatible. $zx + zx^2$ combined, I thought, into $2zx^3$, although if I had thought harder I would have realized that I had added $z + z$ and multiplied x^2 by x, meaning I had been inconsistent. One term is $(zx)(1)$ and the other is $(zx)(x)$, so they were not like terms at all.

"They have to be exactly the same to combine?"

"That's right."

"So, you mean it's actually literal?"

"That's the beauty of it," she said. "It's not mysterious. It's not magic at all."

This was frustrating for me, even unsettling, even anger-making, because the procedures had seemed *not* literal, *not* reliable, and actually magical. The rules I thought I could depend on had failed me, because I had understood them imperfectly and applied them inconsistently, just as Poincaré had predicted that saps like me would do.

"For the moment, think of it as a monastic discipline," Amie said. "You have to take on faith what I tell you."

AMIE AND I have always been close and I could see that she wanted to help me, but she couldn't always see how to. I think she might agree with Poincaré that mathematics was a question of logical remarks following one another in an orderly way, and that this was especially true of mathematics as rudimentary as the version I was struggling with. I was compromised, clearly, by my own shortcomings, but not in a way that her experience might help her understand. My habits of mind were simply different from hers. She didn't have the same shortcomings, and so far as amateur mathematics was concerned, she probably had no shortcomings at all.

As the math grew more difficult, Amie's explanations tended to become either too complex for me or too opaque. Sometimes what she told me made sense when I was talking to her, but I couldn't repeat it on my own. I might recall most of what she had said, often I had written it down, but the explanation relied on procedures I hadn't understood sufficiently to reenact. Now and then I could hear in her voice an exasperation from her not understanding why I couldn't seem to grasp the simplest concepts. It is one thing to teach someone what appears to be a straightforward discipline and another to understand why it doesn't seem straightforward to him or her. One process is procedural, and the other requires a sympathy of imagination that has little to do with being a mathematician. For Amie mathematics was logical, and for me it wasn't. She saw patterns where I saw chaos, incoherence, obfuscation, and conspiracy.

14.

When I told Amie how much trouble I had with math as a boy she said, "You were probably taught badly." I have no idea if the men and women who taught me and my friends were good at it or had any enthusiasm for it, either. They had turned the pages of the same textbooks year after year, and, so far as they knew the material, could probably have taught math in their sleep. They drove old cars and took second jobs in the summer, and some of them were likely in unhappy marriages or drank or were lonely and felt that everything that was ever going to happen to them had already happened and who knows what was on their minds as they looked out at a classroom of adolescents, only a few of whom appeared to be paying attention, and recited problems they'd been reciting for years. I'm not comfortable finding fault with them. If I were to return to algebra class, I would approach Mr. Carmine Biazzo's desk on the first occasion where I had difficulty and ask to have the matter explained again. He might just want to get on with his day and tell me that he was busy. Or he might put everything else aside and try to address my confusion. There is also the chance that I might still not get it, and he might conclude that in trying to help me he was spending his time unwisely.

Learning is a form of adaptation and of receptivity.

Learning math more complicated than arithmetic means absorbing and remembering a wagonload of information and then using it to reason. Powerful impressions are more likely to last. As people get older, though, especially older than sixty-five, their ability to collect new memories diminishes. Learning also takes longer. Moreover, creative thinking is compromised, since the mind is inclined to follow patterns it knows. This may be a reason why pure mathematics is regarded traditionally as a young person's pursuit.

The ability to learn mathematics is thought to decline around forty, when the brain begins slowing its handling of procedural operations such as calculating. Older people learn and forget at roughly the same pace that younger people do, but calculating takes an older person twice as long. In the paper "Acquiring Skill at Mental Calculation in Adulthood," Neil Charness and Jamie Campbell say that middle-aged people perform as older ones do, but if they practice, they perform more as younger people do. If speed is valued more than accuracy, the decline in ability is obvious. If accuracy is valued more than speed, the decline is less obvious and maybe not even very pronounced. Younger people tend to read faster than older people. Older people tend to remember more of what they've read.

From brain scans it appears that older people engage more of their faculties in solving a problem than young people do. Older brains might be less robust, but they may also have become more efficient. The Scaffolding Theory of Aging and Cognition says that brains respond to declines by recruiting assistance—that is, by replacing a response typically dedicated to a single area with a pattern of layered responses involving several areas. "HAROLD" is an acronym for "hemispheric asymmetry reduction in old adults," a form of brain plasticity. I know about it from the research article "Creativity and Aging," by Gene Cohen. It means that brain activity in older people tends to be "less lateralized" than in younger people, Cohen says, meaning that the brains of older people might enlist areas that usually have one function to collaborate with another function, which is called bilateralization. Cohen likens it to the brain's moving, perhaps in a compensatory way, "to an all-wheel drive." A study at the University of Toronto found that older people did as well as younger ones on visual tests relying on short-term memory, but the areas of the brain that the younger people used were weaker in the older people, so the older people engaged other areas. One of the other areas was the hippocampus, which would more usually be invoked for a task such as learning a long speech.

Dr. Carol D. Ryff, at the University of Wisconsin's Institute of Aging, told me about stereotype embodiment theory, which was proposed by the Yale psychologist Becca Levy. It says that the culture presents older people as moving slowly, being hard of hearing, talking too loud, and unable to read small print. These depictions are funny when we're young; then we grow old and enact them, and they undermine a person's sense of well-being. "There are certain fields where you get better with age, though," Dr. Ryff told me. "You're not going to have a twenty-two-year-old wunderkind psychotherapist. Most of Freud's brilliant theories didn't arrive until his fifties."

I told Dr. Ryff that I was trying to learn math, and that I had a math allergy. "Someone with math anxiety, later in life, with a different perspective can really shine and discover something new," Dr. Ryff said. "It's incredibly healthy for the brain as well." I wasn't completely confident that I had yet developed a new perspective, so I was less cheered than I might have been.

I do not think as fast as I used to, and I do not think by means of the same associations and patterns. I do not mind this. Thinking quickly often meant that I responded impulsively and made difficulty for myself. I used to believe that such behavior was evidence of my having a passionate nature. Also, that it was winning. Now I just think I was an idiot. Anyway, I am no longer so prone to irresponsible remarks, and I am grateful for having fewer rough edges.

Sometimes I have to wait for words to arrive, but I seem also to have handled this in a manner that is suggested by some of the brain studies. Not long ago I found myself trying to recall the name of a friend's cat. What I came up with, Leonard, was not close. My mind supplied a generic image of a dog, then a German shepherd, then a German shepherd that belonged to an older couple I knew who lived in my friend's town. I heard the dog's owner, a cultivated old European man named Serge Chermayeff, calling the dog: "Myyyyyyyylllllllowwww." The cat's name was Milo. This took about three seconds.

When I turned sixty-five, I thought, I am as far from fifty as I am from eighty, which sobered me right up.

15.

My capacity for slower thinking, for holding an idea longer in mind, and treating it more deeply, my ability to do this has grown, as I might have hoped when I was younger, if I had known that such a thing was possible. A lot of things about getting older are beyond our imagining when we are younger, or understanding, even if they were made clear. I had no idea when I was young of the wealth and depth of experience and learning and sometimes wisdom that an older person brings to a conversation. I wince when I think of how little of other people's lives I understood when I was young, especially the lives of older people. I tended to think of older people as having always been who they were at the moment I was talking to them. I lived almost entirely within myself, which I think is a defective way of being a person, considering all the freedom, the enlargement of oneself, that a penetrating sympathy makes possible.

Early in my studying algebra for the second time, I found I could learn things sufficiently to employ them in an exercise, but not sufficiently to remember them when they reappeared in a different context weeks later, say. It occurred to me that I wasn't really doing math; I was doing what in my childhood was called learning by rote. In the journal *Science*, I read of a study by researchers in France and at MIT saying that learning multiplication tables is more like memorizing a laundry list than it is doing math.

Learning a language, one might forget a word and do without it, there are other words, but the vocabulary of mathematics has very few synonyms. The difference in being older was that unlike in high school, when I saw no relations among fields or methods, when the numbers and equations almost seemed to blur on the page, now I (slightly) more often understood what is connected to

what and how, and usually I can figure out where to find
a solution. I wish I had been able to do this when I was
young, but I hadn't.

16.

I have deeply regretted that I did not proceed far enough
at least to understand something of the great leading
principles of mathematics, for men thus endowed seem
to have an extra sense.

 —Charles Darwin,
  The Autobiography of Charles Darwin

My suspicion that by not learning adolescent mathematics, my ability to think expansively might have been
impeded seems to be supported by the study "Origins of
the Brain Networks for Advanced Mathematics in Expert
Mathematicians," by Marie Amalric and Stanislas Dehaene. Some scientists think that the areas of the brain
that handle mathematics are ones involved principally
in language, and some think they are ones that handle
number thinking and spatial reasoning. The language
endorsers believe that language developed first and that
numbers and mathematics developed as a consequence.

 Scanning the brains of fifteen mathematicians and fifteen non-mathematicians while they considered complicated mathematics and also questions mostly involving
history, Amalric and Dehaene found that in the mathematicians and not in the others the regions involved in
considering math problems were separate from those

that make sense of language. Language operations take place mainly in the left hemisphere. The problems that the mathematicians considered, which were in algebra, geometry, analysis, and topology, used areas in the front and middle of the brain that are engaged in thinking about space and number.

The part of the brain that performs precise calculations is different from the one that estimates them. An earlier study that Dehaene was involved in found that the calculating part used an area of the brain typically invoked when remembering. The approximating, though, took place among a network of areas participating in "visual, spatial and analogical mental transformations." By "transformations" I assume they meant stages of reasoning.

They concluded that the mathematical areas were associated with "an evolutionary knowledge of number and space." This might be because mathematical insights sometimes are worked out as forms of approximation, which is sympathetic to intuition and involves spatial concepts such as the number line, a thing a person tends to see in the mind's eye. Their conclusion might have been endorsed by Einstein, who said, "Words and language, whether written or spoken, do not seem to play any part in my thought process."

Amalric and Dehaene believe that ease with space and number in childhood might be a reliable predictor of how well someone will do with math. I sometimes walked into walls as a child, and one of my adolescent friends told me, "You have no spatial relationships," which I took as a rebuke.

It turns out that if we believe we can learn, we do better than if we don't. After my first encounter with

algebra, it had never occurred to me again that I could
learn math.

17.

Amie thought I should read *How to Solve It*, by George
Pólya, a handbook for mathematical problem-solving which
was published in 1945 and was part of her reading as a
freshman at Harvard. Mathematics, Pólya writes, has two
identities. As a field it is a "systematic deductive science." As
an endeavor, it is "an experimental inductive science."
Classically, it involves two types of reasoning, analytic
and synthetic. One moves forward and one backward. To
prove A equal to E, synthetic reasoning establishes that
A is equal to B, B to C and C to D and D to E. Seeking to
equate A to E, analytic reasoning likens E to D and D to
C and so on. Analysis forms a plan that synthesis enacts.

In *Discourse on the Method of Rightly Conducting the Reason, and Seeking Truth in the Sciences*, which was published
in 1637, Descartes describes practices he established for
himself for finding solutions. In old-fashioned prose so
full of life that he seems nearly present, as if testifying, he
writes, "I was then in Germany attracted thither by the
wars in that country, which have not yet been brought to
a termination; and as I was returning to the army from
the coronation of the Emperor, the setting in of winter
arrested me in a locality where, as I found no society to
interest me, and was besides fortunately undisturbed by
any cares or passions, I remained the whole day in seclusion, with full opportunity to occupy my attention with
my own thoughts."

Descartes arrives at four precepts that "would prove perfectly sufficient for me, provided I took the firm and unwavering resolution never in a single instance to fail in observing them." They amount to a kind of diagram for how to think. He writes:

> The *first* was never to accept anything for true which I did not clearly know to be such . . . to comprise nothing more in my judgment than what was presented to my mind so clearly and distinctly as to exclude all ground of doubt.
>
> The *second*, to divide each of the difficulties under examination into as many parts as possible, and as might be necessary for its adequate solution.
>
> The *third*, to conduct my thoughts in such order that, by commencing with objects the simplest and easiest to know, I might ascend by little and little, and, as it were, step by step, to the knowledge of the more complex; assigning in thought a certain order even to those objects which in their own nature do not stand in a relation of antecedence and sequence.
>
> And the *last,* in every case to make enumerations so complete, and reviews so general, that I might be assured that nothing was omitted.

After the manner of Descartes, Pólya organizes problem-solving according to four phases. The first requires understanding a problem, so as not to be undermined by misguided action. The second involves planning an approach that one carries out in phase three. In phase four a person examines the problem for any lessons it provides.

Pólya amends each phase. In assessing a problem it is important to consider whether one knows similar problems with similar unknowns and can borrow tactics or find a simpler version of the problem to solve. Straying too far from a problem risks losing sight of it altogether, though. Borrowing from Descartes, Pólya says that the best practice is to state the problem and then separate it into parts. A way out of being stalled is to identify the parts that are difficult and then look for similar examples that have solutions. One could also try putting the parts in a different order. "Difficult problems demand hidden, exceptional, original combinations, and the ingenuity of the problem-solver shows itself in the originality of the combination," Pólya writes.

No idea is bad unless a person is uncritical. Accepting a guess as a truth, as superstitious people do, is misguided, but so is ignoring a guess, as pedantic people do. As regards ideas, it is only bad not to have any.

Resolve matters, too. "It would be a mistake to think that solving problems is a purely 'intellectual affair,'" Pólya writes. "Determination and emotions play an important role. Lukewarm determination and sleepy consent to do a little something may be enough for a routine problem in the classroom. But, to solve a serious scientific problem"—or a pressing one or to accomplish a serious task, I can't help thinking—"will power is needed that can outlast years of toil and bitter disappointments."

Figures farther back than Plato stand behind Pólya's assertion that "teaching to solve problems is an education of the will," and that a rigorous education imposes character. I read Pólya closely, not only because Amie had recommended it, but also because, while I wasn't moving toward a future in which I was necessarily going to solve complicated math problems, I was surely, in growing older, moving toward one with complicated problems.

18.

Something in me gave way, slowly slowly, against the conclusion that when I got a wrong answer, I was more likely wrong than math was. I had railed against poor writing, withheld information, apparently illogical procedures, the clannish perversity of mathematical reasoning, basically anything I'd encountered that hadn't seemed clear as water, and I hadn't won a single round, except maybe against poor writing. The wear and tear was wearying, and I knew Amie didn't want to hear me rant any longer. I decided to try to find what pleasures math had if I didn't always fight it.

I tried to study things objectively rather than examine them for hidden information intended to mock and deceive me. I am aware that many people would not feel aggrieved the way that I did and would begin their studies less combatively, but those weren't my circumstances.

Now and then I backslid, usually on occasions when the rules weren't obvious or easily intuited or seemed arbitrary, and then I'd get angry and have to have Amie calm me down. Studying factorization, for example, I learned that 2^{-2} is not 2×-2, which is -4. By the rules of exponents, Amie said, 2^{-2} is $1/2^2$, which is $1/4$. The discussion over why this was so was long and contentious, and, although my objections were cunning, I lost.

"All the rules are absolutely logical. Please don't say anymore that there are contradictions in math," Amie told me. "There aren't."

19.

About contradictions: In the sciences other than mathematics, the most authoritative finding is the result of observation and deduction and the acceptance of things that appear to be true pretty much all of the time. In mathematics, the paramount authority is proof, and a single exception among infinite cases disqualifies a result. "I'm almost certainly right" is for disciplines where the result can be argued.

A mathematician's engagement with a problem resembles an artist's, but is different in that an artist is not confined to a single outcome. Hemingway wrote forty-seven endings to *A Farewell to Arms*. A math problem typically has a single solution no matter how many ways a mathematician tries to approach it. In *Love and Math*, the mathematician Edward Frenkel writes, "With math you know what you're trying to solve and either you do or you don't." Unlike an artist, a mathematician has the satisfaction of knowing that no revision will change an outcome and that he or she is right for all time and in all circumstances. According to Reuben Hersh in *The Mathematical Experience*, a mathematical proof aspires to be "so bound up with what is right in the universe, that God almighty could not set it aside." This makes mathematics different from other arts in that the best mathematics is permanent. Even though W. H. Auden, in "The Dyer's Hand," writes, "The whole aim of a poet, or any other kind of artist, is to produce something which is complete and will endure without change," decisions about permanence in the other arts are only agreements and revocable. Cultural assertions and their revisions, no matter how emphatic, reflect only who we believe ourselves to be at any moment. Cultures have conscious and unconscious lives, too.

An artist in another field must also always wonder what other form a work might have taken and whether that form might have succeeded better than the one that he or she chose. A mathematical solution is absolute, although in *How Not to Be Wrong*, Jordan Ellenberg points out that a lifetime might not be sufficient to achieve it. "Fermat's problem took 350 years," he writes. (Fermat's problem, usually called Fermat's last theorem: no three positive integers A, B, and C satisfy the equation $A^n + B^n = C^n$ if the value for n is a whole number greater than 2.)

SOME PEOPLE HAVE a talent for obliquely penetrating the concealed design of complex math problems. Where the approaches of most mathematicians appear straightforward, these others digress, sometimes eccentrically. The mathematician Goro Shimura often collaborated in the 1950s with Yutaka Taniyama. About Taniyama, in *The Map of My Life*, Shimura writes that "though he was by no means a sloppy type, he was gifted with the special capability of making many mistakes, mostly in the right direction. I envied him for this, and tried in vain to imitate him, but found it quite difficult to make good mistakes."

Models for proofs descend from Euclid. In its earliest history mathematics involved observations, inferences, and superstitions about numbers, but after Euclid it was deductive. In *The Music of the Primes*, Marcus du Sautoy likens this change to the one when alchemy became chemistry in the seventeenth century.

Mathematical proofs typically use the law of the excluded middle, meaning that they are either/or. Most proofs are straightforwardly deductive, but mathematics also uses an indirect method, borrowed from logic, called proof by contradiction, which is sufficiently well known that I probably don't need to add that it is also called "reductio ad absurdum." This method proves a statement true by assuming it is false and then showing that such a conclusion produces a contradiction. Euclid's proof about prime numbers, those numbers such as 2, 3, 5, 7, 11, 13, and so on that can be divided only by themselves and 1 without leaving a remainder, is the first proof of there being an endless number of primes. It is also a proof by contradiction.

Euclid supposes that there is a largest prime, P. Then he defines a number Q by multiplying all the primes and adding one, in other words that $Q = (2 \times 3 \times 5 \times 7 \ldots \times P) + 1$. Q is not prime, because, by the terms of Euclid's argument, P is the largest prime. Divided by any of these prime numbers, though, Q will leave a remainder of 1. If Q is not a prime, then there is a prime that can divide it, which would be a prime greater than any on the list (and might be Q itself). This contradicts the notion that there is no greater prime than P, so the assumption of there being a final prime is false. This is fairly clear-cut, but I had to go through it with Amie so many times that I no longer remember why I found it so hard.

Euclid didn't invent the concept of proof so much as systematize it. The concept was introduced in the sixth century BCE by Thales and by Pythagoras as a means of establishing assertions that weren't obviously true. According to Ian Hacking, in *Why Is There Philosophy of Mathematics At All?*, this appears to be the result of the Greek culture's being essentially argumentative, whereas cultures in China and Mesopotamia, which also had sophisticated mathematics, were authoritarian. Hacking refers to an observation made by Geoffrey Lloyd, the British historian of ancient science and medicine at Cambridge, that "the hierarchical structure of a powerful education system, with the Emperor's civil service as the ultimate court of appeal, had no need of proofs to settle anything."

A MATHEMATICAL PROOF is a faultless argument establishing a statement from a preceding one or from an axiom, which is a remark so obvious that it doesn't need a proof: "between any two points a single straight line can be drawn" is an axiom. A successful proof is a theorem. A statement in mathematics that has no proof is a conjecture, as in the Goldbach conjecture, the most famous unsolved problem in mathematics. The Goldbach conjecture says that every even whole number greater than 2 is the sum of two primes. It has been found to hold as far as 400 trillion, but a number might exist that refutes it, so it's a conjecture.

Propositions, lemmas, and corollaries also have proofs. Theorems may be considered to be very important propositions. Lemmas are minor propositions that lead toward a theorem, and corollaries are propositions that follow from a theorem. These terms are subjective, though, and deciding when to call a result a lemma or a corollary rather than a theorem or even a proposition is a matter of a mathematician's judgment.

Descartes believed that the best proofs could be understood at one reading. Leibniz thought that the best proofs advanced toward a result that couldn't be grasped all at once. For Bertrand Russell the appeal of a proof is not only its result, as people tended to think, but also the elegance of its structure. "An argument which serves only to prove a conclusion is like a story subordinated to some moral which it is meant to teach," he writes. "For aesthetic perfection no part of the whole should be merely a means." Impatience, he thought, led people to overvalue a proof's result at the expense of its claims.

AMONG THE MOST useful methods for arriving at a proof, applicable to problem-solving in general, is to work fluidly, that is, forward and backward. In "A Brief Introduction to Proofs," William Turner writes, "Start with your hypothesis and ask yourself what does this imply, but also look at your conclusion and ask what you need to prove to get to it." Sometimes a proof by contradiction is easier, but "if you have trouble proving it false, try to use the reason you are running into difficulties to prove it true." Turner likens a proof to a map leading from a hypothesis to a result. Whether it was built consecutively or from both ends to the middle matters less than whether it is correct.

Mathematics has two pursuits: to find patterns and to prove that the patterns are lasting ones. Before the seventeenth century mathematics was more a component in other sciences than a science itself. If a piece of mathematics was useful, it didn't necessarily require a validation for why it worked. The modern figure influential in changing this was the German mathematician Carl Friedrich Gauss, who believed that mathematics must be justified and that proving things is what a mathematician should do.

A work of art rewards repeated engagements. For this reason, mathematicians like to return to certain proofs and mathematical documents. Gauss said, "You have no idea how much poetry there is in a table of logarithms."

20.

I passed the first weeks of my sojourn in the algebra highlands learning rules such as the commutative law ($a + b = b + a$; $ab = ba$) and the distributive law ($c \times (a + b) = c \times a + c \times b$)). I had also to learn the properties of fractions; how to add and multiply negative numbers; the terms governing powers and negative powers and the multiplication of powers; and how to manage polynomials, the expressions that have letters and numbers, such as ($a + b$) ($a + 2b$), and so on. As a boy my eyes glazed and they still do when I see formulas and equations, especially ones I haven't seen before and especially ones that I don't understand. They provoke a kind of weariness of the soul. Even so, formulas and equations were not the reason I had failed to learn algebra. The reason I had failed to learn algebra was word problems. From the first day of this endeavor, opening the first textbook, I worried about word problems. I recalled them as devious and irrational, the evil end of the language ladder whose higher end was poetry.

Gustave Flaubert sent his sister Caroline, who was studying mathematics, a word problem in 1841. "A ship sails the ocean," he wrote. "It left Boston with a cargo of wool. It grosses 200 tons. It is bound for Le Havre. The mainmast is broken, the cabin boy is on deck, there are 12 passengers aboard, the wind is blowing East-North-East, the clock points to a quarter past three in the afternoon. It is the month of May. How old is the captain?"

I found this problem in *Mathematics and the Imagination*, by Edward Kasner and James Newman, which was published in 1940, and if they hadn't written that it "cannot be answered even though there seems to be plenty of information supplied," I would have tried. I regarded all word problems as illogical, so I wouldn't have seen Flaubert's as being any more illogical than this one, from the Khan Academy website: André, a collector, bought 4 baseball cards. The next day, a Tuesday in March, on which it was raining, his dog ate 50 percent of his collection, leaving André 20 cards. André's father is a doctor, and his mother is a biology professor. They gave André the cards to start his collection. How many cards did André's mother and father give him? Which does have an answer: 36. ($\frac{1}{2}$ (x + 4) = 20).

Another numbing word problem, a relic one, from *A History of Pi*, by Petr Beckmann, provided by "The able Chinese mathematician Sun-Tsu (probably 1st century A.D.): A pregnant woman, who is 29 years of age, is expected to give birth to a child in the 9th month of the year. Which shall be her child, a son or a daughter?"

Sun-Tsu's solution: "Take 49; add the month of her child-bearing; subtract her age. From what remains, subtract the heaven 1, subtract the earth 2, subtract the man 3, subtract the four seasons 4, subtract the five elements 5, subtract the six laws 6, subtract the seven stars 7, subtract the eight winds 8, subtract the nine provinces 9. If the remainder be odd, the child shall be a son; and if even, a daughter." Occasions such as these reinforce my sense of mathematics, in its nature, as a series of questionable propositions.

Amie did not endorse my studying word problems. She said that they taught nothing of any lasting use and, once disposed of, never appeared in mathematics again, an argument I wish I had been equipped to present to Mr. Carmine Biazzo. I understood that they were a byway to her, but in the math curriculum, beginner division, they are prominent, at least as I remember it. Perhaps this is because ancient algebra begins with word problems. The earliest document with algebra on it is the Ahmes Papyrus, an Egyptian scroll in the British Museum that was copied by a scribe named Ahmes "in the year 33, in the 4th month of the inundation season." Scholars believe that this might be around 1650 BCE. The scroll describes itself as "the entrance into the knowledge of all existing things and all obscure secrets," and it is thought to be a math textbook. In addition to tables concerning addition and multiplication, instruction on how to handle fractions, geometric definitions, and methods for finding areas and volume, it has eighty-four problems, one of which, the twenty-fourth, an example of rhetorical algebra, begins, "A heap and its 1/7 part become 19. What is the heap?" According to the translation made by the Mathematical Association of America in 1927, "The author assumes 7, which with its 1/7 makes 8, and then, to find the answer, he multiplies 7 by the number that multiplying 8 will give 19." In *Algebra for Dummies* I find that $x + x/7 = 19$ works, too.

When I began this project, I knew like I knew my name that I couldn't do word problems. Reading one, I would hear a companion voice say, "No way you're solving this. Not you, not now. *Flintstones, meet the Flintstones, they're* . . . Fine, ignore me. I'll just tag along, shall I?" Reading a word problem, I would have the sensation that I was sliding over the surface of the text, as if in a car that was skidding on black ice. I could find no place to stand from which I could say confidently, "This is the information I need to solve this problem, this is how it goes together, and I can ignore the rest."

The other children seemed already to know what to do. So who had taught them? Why had I been left out? I say this to demonstrate how far I had wandered from the simple and straightforward. Word problems were both a practical and a metaphysical problem for me. They were a species of taunt and the sneering face of trouble in the mirror.

"Word problems aren't the same as math problems," Amie said sympathetically. "They lie in some space between reading and math, and they can really fell people who have issues with reading and attention. So much intervenes in the head of a twelve-year-old who isn't, or is not yet able, to grasp complicated sentences. If you can't keep all the terms in a paragraph in mind until you reach the end, you're in trouble." I decided not to say that I wasn't twelve.

Words in word problems don't behave as they do in conversation or as words generally do on the page. Instead, they have a background meaning; they conceal the means of solving the problem, as if they were encrypted. On the surface they are a statement of circumstances—a traveler, a train, Topeka and Dallas, say—but they are also vessels of meaning for numbers and relations of consequence. Prose writing typically has an explicit meaning. It might suggest other meanings, but it intends to convey a thought or a series of thoughts. Word problems use language subversively. The car, if there is one, isn't actually a car. It is irrelevant to ask if the car is a sedan or a wagon, a Ford or a Chevy, new or secondhand—information that might help you picture it—because the car stands for velocity. It isn't a metaphor, it is a car being used as a proxy for an object moving in a specified direction with reference to time.

I was an imaginative child, that wasn't my problem. My problem was that no one had explained to me the steps of inquiry involved or even that there were steps of inquiry. Word problems embody a type of dubious abstraction that algebra thrives on. I am not the only one who has noticed this. In *Ulysses*, a remark by Buck Mulligan leads the Englishman Haines to ask Stephen Dedalus his idea of Hamlet. Haines and Buck Mulligan have the following exchange:

> You pique my curiosity, Haines said amiably. Is it some paradox?
>
> Pooh! Buck Mulligan said. We have grown out of Wilde and paradoxes. It's quite simple. He proves by algebra that Hamlet's grandson is Shakespeare's grandfather and that he himself is the ghost of his own father.

21.

I read Amie a simple problem, "Cletus is driving from Austin to Houston, the total distance is 162 miles. There are 5,280 feet in a mile. How many hours will it take

Cletus to get to Houston if he drives at an average rate of 88 feet per second?"

"This is a good one," she said, "because it's not written like a real series of sentences. In real life you expect each sentence to comment on the previous sentences, but here you're handed three unconnected ones."

"What would you do?"

"I'd think, Where do I start? I would always draw a picture."

Drawing a line on a piece of paper, she said, "I'd put Cletus at one end. First I multiply by 60 because there's sixty minutes in an hour, then I multiply by 60 again, because 60 seconds in a minute. So $3600x$ is the number of seconds. $3600 \times 88 \times x$ divided by 5280 is the number of miles he would travel, which is 162. We want him to travel that distance in x hours . . ."

In school every word problem became an equation of the $5/12 = x/20$ type. Eventually Amie arrived at 2.7 hours, but since she didn't write such an equation, and I couldn't adapt her words to the form, I was still as lost as I could be.

22.

If I couldn't solve word problems, I didn't feel I could say I had mastered, or at least passed, algebra. Doing them was like visiting a place where something unhappy had occurred and hoping that I might be able to shed my disabling attachment to it. Also, if I couldn't get past them, I didn't see how I could go on to geometry and calculus with any confidence.

As a boy, I tended to read the problems too quickly, feeling that the sooner I got to the end the smarter I was, and if I didn't have an equation in mind, I sulked. I had a magic-thinking notion that if I knew what I was doing the equation would reveal itself, as if written in invisible ink.

Reading a problem a second or third time wasn't going to tell me anything I didn't already know. And anyway the words would blur again. A car was traveling between two places I'd never been. How the car related to the time, or the distance to the place, I couldn't tell, even though I had sat in a classroom where the reasoning had been demonstrated plenty of times. I don't know why I hadn't been able to absorb it. I just hadn't. In a way that no other subject is, except perhaps physics, math is indifferent to the well-being of those who fall behind. It throws off stragglers remorselessly.

In my second engagement, I read problems carefully, methodically even, like a particularly dim detective, doubting everything. I should be able to do this, I thought. It's not like it's magic. I continued to believe, though, that they were deceptive. Here is how I would go wrong: Jane is 20 years younger than Rebecca. Rebecca and Jane first met 2 years ago. 14 years ago, Rebecca was 3 times as old as Jane. How old is Rebecca now?

Begin with Rebecca, I thought, who is 20 years older than Jane, so, to give her current age a variable, she is $R = J(ane) + 20$. Then I thought, I also have the matter of 14 years, so I wrote another equation saying $14Y = 3R = J + 20$. They don't go together, because they can't. They are useless. And I wince at having thought of them, let alone for congratulating myself at dismissing the superfluous mention of their meeting two years earlier.

The solution perhaps is obvious; at least it was for me after Amie explained it. Jane, being twenty years younger than Rebecca, is $J = R - 20$. Fourteen years ago, $R - 14$, Rebecca was three times as old as Jane, $R - 14 = 3(J - 14)$. Since $R - 20 = J$, it can be substituted into the equation in place of J, in order to eliminate the second variable, therefore $R - 14 = 3(R - 20 - 14)$.

$R - 14 = 3(R - 34)$. For math greenhorns like me, I will explain, so as not to skip steps, that negative numbers added to each other produce smaller negative numbers; smaller being a matter of greater distance on the number line from zero. If I begin at -20 and take 14 steps in the negative direction on the number line, I arrive at -34, a smaller number than either -20 or -14.

$3 \times R$ becomes $3R$ and 3×-34 becomes -102, thus,

$$R - 14 = 3R - 102$$

Anything done to one side of an equation must be done to the other, so to consolidate the terms I add 102 to both sides, $102 + (-102) = 0$ and $102 + (-14) = 88$.

$$R + 88 = 3R$$

Subtracting R from the left-hand side of the equation leaves 88, and subtracting it from 3R leaves 2R.

$$2R = 88$$

$$R = 44$$

Fourteen years ago, Rebecca was 30 and Jane was 10.

I didn't expect to become accomplished at word problems, but I didn't expect to have them trounce me again, either. I refused to give up on them until I felt at least modestly competent. If I could get four out of five correct,

I felt that would be sufficient. Instead, I often went something like one or two for five. After I finished the problems in the textbook, I did them on a website that had an inventory of them, but I worked so doggedly that I used up their archive and eventually I was offered problems I recognized. Occasionally I would have forgotten the solution and could work them again, but after a while it was like seeing a movie so many times that you can repeat the dialogue.

23.

Learning algebra, a person is crossing territory on which footprints have been left since antiquity. Some of the trails have been made by distinguished figures, but the bulk of them have been left by ordinary people such as me. As a boy, trying to follow a path in a failing light, I never saw the mysteries I was moving among, but on my second pass I began to. Nothing had changed about algebra, but I had changed. The person I had become was someone whom I couldn't have imagined as an adolescent. Math was different, because I was different.

The first mystery, waiting like a welcoming party in the vestibule of mathematics, concerned the origin of numbers. It's a simple speculation, where do numbers come from? They don't typically appear in creation stories. In *Chinese Myths*, Anne Birrell writes that the Coiled Antiquity myth, which belongs to "a minority ethnic group of south-western China," describes "how numbers were created" and "provides the etiological myth of the

science of mathematics," but she does not give a source
for this assertion, and I have not been able to find one,
so I have to accept her word for it. So far as I am aware
otherwise, no culture has a story where a creature or a
spirit gives numbers to humans. No one in any scripture
I know of climbs a mountain and gets numbers or finds
numbers while wandering in the desert or has a number
dream or a number vision, and I don't know of a figure in
a myth or legend who does, either. No country has a hol-
iday for the day they got numbers. Gods and protectors
of numbers are also rare. Plato says in *Phaedrus* that he
had heard that in Egypt there was a god named Theuth,
"who invented number and calculation, geometry and as-
tronomy, not to speak of draughts and dice," but that is
the only other ancient reference to gods and the creation
of numbers that I have been able to find, and it isn't clear
that Plato didn't make this up. Babylonian, Indian, African, Norse, and Native American myths and traditions,
so far as I can determine, are about other things than
numbers. The people who wrote creation stories probably thought of numbers as practical objects, like the axe
and the wheel, and didn't feel that they required a mythic
explanation.

I don't remember learning numbers any more than
I remember learning to walk or to speak. I think this is
common. As very small children, most human beings can
look at a small collection and say how many objects are in
it. This is called subitizing. Pigeons, crows, monkeys, and
dolphins can do it, too. In humans this capacity seems to
fail above seven objects. In *Numbers and the Making of Us*,
though, the anthropologist Caleb Everett describes an
Amazonian tribe who have no words for numbers and

can reckon correctly only with groups of three or fewer, which suggests that recognizing numbers is at least partly something we learn and not an inherent human trait.

Arithmetic begins with combining and condensing. Mathematics begins with measuring, when whole numbers show the ability to become fractions and decimals. Arithmetic is visible. Whole numbers describe objects— three horses, six birds, two bushels of wheat. You can add one collection to another or remove one and compute the result. Mathematics, on the other hand, is largely invisible. You can represent it with symbols, you can discuss it, but you can't necessarily see it. I can imagine an endless series of numbers, but I can't see it. I can see only a metaphor. Arithmetic is a part of actual experience. Mathematics, as James Robert Brown identifies it in *Philosophy of Mathematics*, is a cerebral task. "Deriving the $\sqrt{2}$ has nothing to do with sense experience," he says.

COUNTING LEADS TO adding and subtracting and the fancier operations of multiplying and dividing. Invoking them at first, are human beings inventing mechanical procedures or observing properties inherent in numbers? As soon as numbers appear are they already intractable? They weren't intended to be, any more than they were intended not to be. At first, they assess obvious and near-at-hand qualities of the natural world, but thousands of years later they demonstrate the ability to describe grander aspects of nature intimately, the orbits of the planets, for example. People used numbers for centuries before they even knew there were planets.

Someone who says that human beings created the operations of arithmetic cannot say that we created the results. 2 objects and 2 objects are always 4 objects. We did not say, We are adding 2 objects to 2 objects and for the sake of clarity deciding that the sum will usually be 4 objects. On all occasions, in all universes, the sum will be 4, even if the term that denotes them is not 4. It is a property of their being, an inflexible trait. "Prayer," a poem by Ivan Turgenev, written in 1883, begins, "Whatever a man pray for, he prays for a miracle. Every prayer reduces to this: 'Great God, grant that twice two be not four.'"

NUMBER SCHOLARS TEND to think that counting was discovered by different cultures at different times, but Abraham Seidenberg, writing in "The Ritual of Counting," published in the Archive for History of Exact Sciences in 1962, thought that "various peculiar features of counting practices" shared among many cultures meant that counting began in one place and spread. Seidenberg thought that counting arose from creation rituals in which the number of people taking part was specified by the ritual. It might have happened that a man entered the ritual first, an explanation for the near ubiquity in ancient cultures of odd numbers being thought of as male and even numbers as female.

Seidenberg says that older cultures sometimes had counting prohibitions and specialty practices for reckoning objects. In parts of Africa, counting a person was believed to bring the person to the notice of the gods, who might decide that the person needed more suffering. Among certain African people who disliked using words for numbers, a speaker would say a number's first syllable and make a gesture and the person he or she was speaking to would say the rest of the word, to spread the risk.

In Oran, in Algeria, counting grain required a counter in a state of "ceremonial purity." The counter would say "'In the name of God' for 'one'; 'two blessings' for 'two'; 'hospitality of the Prophet' for 'three'; 'we shall gain, please God' for 'four'; 'in the eye of the Devil' for 'five'; 'in the eye of his son' for 'six'; 'it is God who gives us our fill' for 'seven'; and so on up to 'twelve,' for which the expression is 'the perfection of God.'"

I find this mixture of ceremony and superstition very moving and sensible, acknowledging, as it does, the fragility of human circumstances, the wish for protection, and our innate belief in magic and the supernatural—in a shared human heritage, that is, and all that lies past what we can be sure of.

24.

It seems not unreasonable to speculate that numbers are latent in the design of the world, but latent where? Think of three animals in an ancient forest. Two humans are observing them. One of them identifies the group, for the first time, as three. The one who identified them as three did not invent the amount. The concept of three is inherent and independent of its first being named. It was three the day before being identified and the day before that and so on. The observer invented a term for describing a quality, but the quality that made the collection three existed whether it was described or not. The person who had no word for its magnitude would still recognize that the three objects had in common a trait with three other objects.

Is this simpleminded? I hope not. It is the kind of thing that fascinates me.

NUMBERS ARE SIMILAR to language in that no one would say that there weren't thoughts before language, although language appears to make possible a means of naming and ordering thoughts and of having complex thoughts that wasn't possible before language. Likewise, beginning with arithmetic, numbers make more complicated mathematics possible. Mathematics leaves the village of counting like a pilgrim headed for the wilderness, a large part of which consists of imaginary trees.

NUMBERS APPEAR TO have been found more than created. The decision whether to call such a magnitude *four* or *quarto* or *quatre* is a matter of language. 4 is a symbol outside of language and not subject to it. Counting invoked numbers, but then it is as if a conjuring gesture occurs or an alchemical property is revealed. An accompanying world comes into being that has its own logic, procedures, rules, and circumstances that owe nothing to a human presence. The trick might be said to have been performed by the first person to realize that one more thing could always be added to any collection of things.

Mathematics might partly be defined as the intent to uncover the properties of numbers. It is a complex pursuit that progresses toward no apparent end and that is complicated in ways that we appear to be incapable of imagining entirely. Mathematics seems to have a vitality that we can't account for, either, and to be indifferent to whether we try. It exists whether we think about it or not.

What distinguishes mathematics among the imaginative arts is its precision. A mathematician might think of any possibility, but what he or she thinks of has to be proved to exist. In mathematics some possibilities exist, others don't. There is no such thing as conceptual mathematics like there is conceptual art. You can say, as Douglas Heubler did in 1970, that you intend to take a photograph of every person on earth to document his or her existence, and you can set about trying and run out of time, but you can't invoke an actual world where the square root of 4 is 0. You can only pretend there is such a world.

MATHEMATICS IMMEDIATELY DEVELOPS two identities, two sides, two aspects, two natures, divides into yin and yang, night and day, the practical and the mystical, because as soon as there are numbers there is the question of where they came from. Many mathematicians regard thinking about where numbers and mathematics come from as philosophy and not mathematics, but I don't see how they are so separate. Mathematics is an imaginative practice, and mathematical philosophy is an inquiry, but the objects at the beginnings of both fields, numbers, contain both use and speculation equally. Numbers name things, but they also represent metaphysical concerns. The first aspect is precise, and the second is speculative, and both are ingrained. It is a light/shadow feature of their being. As with a stranger, where are you from doesn't have to be asked, but it's present in the encounter. I can care or not care, have the habit of mind interested in thinking about the mystery or not, but I can't claim that it isn't there.

25.

Numbers have a kind of sway over us, but we have no sway over them. Nothing we do can change what they are or how they behave. They were here before us and will remain if we disappear. They are always past, present, and future, and we cannot be sure that we are. I can say that the Statue of Liberty exists whether I think about it or not, because you and I know where it is and what it looks like. I can also say that the number 4 exists whether I think about it or not, but neither you nor I know where. If I say that it exists only in my mind, or in yours, then where would it be if not a single human mind were thinking of it? It isn't lights out then for numbers.

All objects other than numbers live either in the world of space and time or in the realm of the imagination, which is a precinct of space and time. By the nature of their dual identity, numbers can travel between these places and connect them. In a letter to the physicist Wolfgang Pauli on October 24, 1953, Carl Jung writes of numbers that they "possess that characteristic of the psychoid archetype in classical form—namely, that they are as much inside as outside. Thus, one can never make out whether they have been devised or discovered; as numbers they are inside and as quantity they are outside."

A number would appear to be as simple as a letter—both are serial implements—but letters are literal and numbers have esoteric attributes. If I write the letter A, it is the letter A; it doesn't represent something, it is something. If I write 4, though, it isn't 4 in the sense that A is A. A is concrete, the manifestation of a sound, and 4 is a symbol, a term denoting a collection. It has no scale or identity. It can be four cats or four galaxies. I can write

4, but I can't say that it *is* 4, at least not all the possible
embodiments of 4. I can demonstrate 4 only obliquely, by
gathering 4 things, *AAAA*, for example.

Numbers were invoked by counting, a form of organization. Letters changed speech from something ephemeral into something capable of being preserved, another
form of organization. By means of addition or subtraction
or some other mathematical operation, one number can
deliver us to another, something letters can't do, though,
unless you think that adding letters to one another to spell
a word is similar, which it isn't; you can't divide a word
by a word or a letter by a letter. You can't have half a
letter. Or the square root of a letter. Or 3.65 percent of
a letter. (Only in mathematics is A/B a sensible remark.)
Numbers have two primary incarnations, positive and
negative, but they also have hidden attributes, such as
being prime. By agreement, we can change how words
are spelled, but we can't change arithmetic. We can allow
theater or *theatre*, but $5 + 7$ we can't do anything to at all.

Numbers are a mystery enfolded into ordinary life.
They surround us the way radio waves and dark matter
do, is how I think of it, and, like hurricanes and white
sharks and big cats, they suggest the edges of the inap-
prehensible. Numbers appear to be unambiguous, in that
when I write a number, I can identify properties attached
to it, but except within the context of the logic that numbers embody I can't say for sure why there are properties.

Numbers did not initially provoke wonder or reverence. They did that later, partly because they invoked
notions of infinity and therefore of God, and, after that,
because they appeared to be a language in which nature
could be expressed. And because on examination they
showed themselves to be complex in ways that had noth-

ing to do with what we thought about them. Not being able to settle on an origin story for numbers means that there is no origin story for mathematics, either. It is a question that is always modern.

Even a slight acquaintance with numbers equips a person to consider large mysteries in mathematics. The simplest unsolved problem in mathematics is called the Collatz conjecture. It is named for the German mathematician Lothar Collatz, who introduced it in 1937, but it is likely older than that and is maybe even ancient, meaning that it might have been known to mathematicians in Babylon, China, and India.

The Collatz conjecture says that every number will return to 1 after a simple process: If it is even, divide it by 2. If it is odd, multiply it by 3 and add 1. 5 becomes 16 becomes 8 becomes 4 becomes 2 becomes 1. 9 reaches 1 in 19 steps. The conjecture is known to be true for all numbers with fewer than 19 digits, numbers less, that is, than a quintillion, but no one knows if it holds for an infinite quantity of them.

"Hilbertian optimism," named for David Hilbert, the German mathematician of the late nineteenth and early twentieth centuries, is the belief that every mathematical problem is in principle solvable. In 2010, Jeffrey Lagarias, an American mathematician who is an authority on the Collatz conjecture, described solving it as "completely out of reach of present-day mathematics." The Hungarian mathematician Paul Erdős said a proof was "hopeless. Absolutely hopeless." It is thought possibly to be undecidable.

NUMBERS ARE LIKE wolves. Only with difficulty can they be made to cooperate past any simple and self-interested exchange. A person sharing their company has to discover what they are inclined by their temperaments to do. The best way to get along with them is to let them do what they want. Essentially they're unbending and can't be made to act in a way that violates their natures. The study of their behavior, to extend the analogy, is number theory. Practical math, called applied math, useful in commerce and engineering, is a different creature. Applied math is the dog, compliant and eager to be helpful.

26.

Antiquity: the first figure known to speculate about the origin of numbers was Pythagoras, a pre-Socratic philosopher of the sixth century BCE. Pythagoras is one of the great seer-like/quasi-holy men, at least he is alleged to be. He appears to have been the founder of a manner of living that had strict dietary practices and required five years of silence as a means of learning self-control. I say appears, because hardly anything reliable is known about him. Pythagoreans tended to attach his name to their writings, so knowing what he wrote, or if he wrote anything at all, is complicated. Some scholars think that the absence in the two hundred years following his death of anyone quoting from any of his books suggests that he didn't write any, perhaps from insisting on silence. It is generally accepted that he left no trustworthy record of himself. These days he would be a celebrated mystic/thinker with no internet footprint.

In *Life of Pythagoras*, Iamblichus, an Arabian philosopher who lived about eight hundred years after Pythagoras, says that Pythagoreans viewed Pythagoras as a type of indefinite being. "Of rational beings, one sort is divine, one is human, and another such as Pythagoras," Iamblichus writes. Pythagoreans believed that Pythagoras knew that the soul was immortal and was reincarnated, sometimes into animals, which he might have learned by visiting the underworld. The philosopher and mathematician Dicaearchus, a student of Aristotle, said that Pythagoras knew that after enough time events recur and that nothing is happening for the first time. Other Pythagorean highlights: he killed a poisonous snake by biting him; once while he was crossing a river, the river spoke to him; on the same day, at the same time, he was seen in two places; he had a thigh made of gold, a mark of divinity; he performed ten thousand miracles. Herodotus says that Pythagoras knew a great deal about sacrifices and rituals and that he agreed with the Egyptians that a body should not be buried in wool, which is impure.

In "Pythagoras as a Mathematician," an essay by Leonid Zhmud published in *Historia Mathematica*, Zhmud says that Pythagoras seems to have been responsible for the theory of even and odd numbers, which Zhmud says is "the first example in the theory of numbers." It begins, "The sum of even numbers is even; the sum of an even number of odd numbers is even; the sum of an odd number of odd numbers is odd."

The most famous object with Pythagoras's name on it is the Pythagorean theorem, $a^2 + b^2 = c^2$, where a and b are sides of a right triangle and c is the hypotenuse. Numbers such as 3, 4, and 5 that satisfy this equation are called

Pythagorean triples and were known in Babylon, which Pythagoras visited. His contribution may have been to see their application to geometry.

Pythagoreans believed that the world's design depended on numbers and that nature had a mathematical structure. In *The Theoretic Arithmetic of the Pythagoreans*, published in 1816, Thomas Taylor writes that Pythagoras defined number as "that which prior to all things subsists in a divine intellect, by which and from which all things are coordinated." Iamblichus attributes to Pythagoras the observation that "number is the ruler of forms and ideas, and is the cause of gods and daemons."

The Pythagoreans gave certain numbers qualities. According to Taylor each of these numbers "was philosophically adorned with various attributes ranging from physical to supernatural to mythic to aesthetic to moral." Pythagoreans believed that One, the unity, was divine, and they called it Apollo. Two was audacity, because before all other numbers it separated itself from the One. Furthermore, two stood for opposites and was ignorance, because it was separated, but it also implied wisdom, since ignorance leads to wisdom. According to Nichomacus, a Greek mathematician, in *The Theology of Arithmetic*, a compendium drawing on various sources—in this case, Anatolius, a philosopher and theologian—"The triad, the first odd number, is called perfect by some, because it is the first number to signify the totality—beginning, middle and end." Four, the tetrad, was "the greatest miracle, a God after another manner (than the triad), a manifold, or rather, every divinity." Five, the pentad, was "the privation of strife, and the unconquered, alteration or change of quality, light and justice, and the smallest extremity of vitality." It was also immortal. The Pythagoreans believed that in addition to the nine planets, there was a tenth, the counter-earth, because ten, the decad, was a perfect number—being the sum of 1, 2, 3, and 4, the first numbers, and it was sensible that the heavens would have a perfect design.

After Pythagoras, mathematics is no longer regarded as merely practical, and it is never again separate from philosophical and spiritual speculations or the presumption of abstract qualities. When I read Sextus Empiricus I sense the frustration that ancient thinkers felt when they tried to get numbers to behave according to rules that humans imposed on them. Empiricus lived in the second or third century CE, exactly when isn't known. He was a Pyrrhonian Skeptic, meaning that he believed that every argument had an opposite and equally persuasive argument. Someone who could hold both arguments in mind attained an objectivity that delivered him or her to a state of tranquility. Empiricus writes in *Against the Mathematicians* that numbers don't exist. All numbers are multiples of 1, the monad, he says. Putting two monads together allows the possibilities of subtraction or addition. If one monad is subtracted from the other, a monad no longer exists. Adding one monad to another gets four monads, not two, because the dyad produced by the first operation consists of two single monads, meaning there are four monads altogether. This makes addition impossible, since "the same difficulty will exist in the case of every number, so that owing to this number is nothing."

One morning in the Algebra Hotel, I wrote "Number is nothing" on a piece of paper and taped it to the wall above my desk, a small defiant act.

27.

By week six, I was losing courage, a little. I had gone far enough, and I had foolishly told enough people what I was doing, that there was nothing to do but keep going. One foot in front of the other toward the hills in the distance until I saw the paths and patterns that everyone else saw.

My specialty of overthinking makes things harder. Also, my resistance to learning new ways of thinking. Also, my suspicion, especially with word problems, that I am being deceived. Every now and then, of course, this is true. When I was a boy, math teachers and writers of math textbooks loved trick questions, and to my astonishment they still do. It seems a shabby device, a wan practical joke. The rhythm of their appearing is more or less the same as the rhythm governing my resolutions to stop looking for trick questions and just answer what's apparently being asked, so I keep falling for them.

Sometimes I feel that these problems are designed to exploit a weakness in my thinking, which is that I search too hard for a solution, which means that they have exploited a weakness in my thinking, the tendency to overthink.

It helps to remember the math professor Morris Kline's observation that especially with arithmetic and algebra simple ideas often took thousands of years to arrive at. My slow, even laborious, progress seems less chastening then.

ANSWERING A WORD problem correctly sometimes feels like making a trick shot in pool. Certain mechanics apply generally, but each shot has specificities, and being good at one shot doesn't mean being good at a different one, even if the difference between them is slight.

I wonder if the talents needed to solve word problems are not congenial to someone whose habits of mind are those of a daydreamer, except that I have been given the example of pure mathematicians who seem to live all but entirely in a dream world. Of course, that doesn't mean that they are good at word problems.

28.

I noticed that musicians seem to include mathematics, even if obscurely, in the way that they regard the world, in terms of melody, rhythm, and chord structure. Often they were comfortable as adolescents with math. I thought this was an observation of my own and was disappointed to learn that others had arrived at it first. Then I was further chagrined to learn that there was even a reaction against it, that musicians don't care to be described as being good at math, they think it is reductive. Still, I think it applies. Harmony, the circle of fifths, the modeling of scales and modes, rhythms and odd time signatures all demand an ability to conceive of the world, or one's part of it, as signified by numbers and mathematical intervals and to count. The harmonic structure of a piece of music is a progression of mathematical forms. The scales and modes and chords built from them are a series of mathematical relationships. In the key of C, which has no sharps or flats, the major scale is C, D, E, F, G, A, B, C. The initial triad is C major, built from the first, third, and fifth tones: C-E-G. C-E-G-B is C major seventh. C-E-G-B-D is C major ninth.

C-E-G-B-D-F is C major eleventh, and C-E-G-B-D-F is C major thirteenth. Chords can be inverted: C-E-G-A, which is C major sixth, is also A-C-E-G, A minor seventh (A-C#-E-G# is A major seventh). Chords can also be substituted for one another: C major seventh, C-E-G-B, includes E minor, which is E-G-B. In both classical and improvisatory music, they can be made to resolve in a number of ways that involve the relations among their intervals. What we hear are mathematical arrangements in the form of tone and rhythm. Leibniz somewhere remarks that music is the pleasure the human mind receives from counting without knowing it is counting.

WORKING AS METHODICALLY as an accountant, I was eventually able to solve word problems more often than not. I didn't have a breakthrough, so much as I just sort of wore them down. I managed more than thought my way through them. Still, I was pleased. I felt I had partly erased a flaw in my past.

Winter

1.

My hard feelings toward math began to be mitigated by an awareness of its enigmas. Of course, I am acquainted only with modest enigmas and those only modestly, but they introduced me to a world I had been unaware of, one that was capacious and profound but not so abstruse that I couldn't find the means to appreciate it, even as a tourist. Numbers began to seem as alive in their being as I am in my mind (and nothing very important has happened to me, except in my mind).

The second starter mystery I encountered is that of prime numbers, those that can be divided cleanly only by themselves and by 1. The first primes are 2, 3, 5, 7, 11, 13, 17, 19, and 23. The number 1 is not prime, because, if it were, the fundamental theorem of arithmetic, which is also called the unique factorization theorem, would not hold. A counting number that is not prime is called a composite number. The fundamental theorem of arithmetic says that any composite number can be expressed as a unique product of primes (1 is not a composite number, either). The primes that express a composite number are called its factors, and composite numbers factor in only one arrangement of primes. $2 \times 2 = 4$. $2 \times 3 = 6$. $2 \times$

$2 \times 2 = 8$. $3 \times 3 = 9$ ... $2 \times 3 \times 3 \times 37 = 666$ ... $29 \times 31 = 899$ ... $2 \times 2 \times 2 \times 5 \times 5 \times 5 = 1000$. If 1 were allowed as prime, no series of factors would be unique. $2 \times 2 = 4$, but so does $2 \times 2 \times 1$ and so on.

Prime numbers are where imaginary mathematics begins. They are an example of our discovering properties of numbers, rather than creating them. Idealism is the name of the eighteenth- and nineteenth-century discipline that believed that the mind creates what we know and that there is nothing we can know that the mind hasn't created. G. H. Hardy writes that prime numbers seemed to him the place where Idealism failed: "317 is a prime not because we think so, or because our minds are shaped in one way rather than another," he writes, "but *because it is so*, because mathematical reality is built that way."

2.

With prime numbers I was made aware of the conundrum of how numbers have properties that no one gave them. And that if mathematics is a human creation, how is it that our creation has attributes that we only partly comprehend and are still discovering? And how is it that our minds could create something that they can't entirely understand or contain? Is it possible to create something useful in pedestrian ways that becomes more than it was intended to be, something that exceeds our grasp and control? (AI might prove to be, but AI is dynamic and can act on its own, whereas numbers are inert.) None of the inventors of language imagined novels and poetry, but the possibility of novels and poetry is inherent in the medium. There are no novels or poems without human beings, but there is math and there are prime numbers without us. It would be as if we had invented language and then discovered that certain words had the capacity to talk to each other when we weren't around. In any book we opened there might be conversations taking place that were unrelated to the text. The exchanges in the dictionary might amount to a din. We would know this to be so, but we wouldn't know why, and while we knew many, many words that could talk to each other, not all words could, and we had no way of predicting what words had this ability and no way of finding them, either. All we could do is go through all the possible combinations of the alphabet looking for them, since they seem to appear randomly, and there exists an infinite collection of them.

Becoming aware that prime numbers were lurking among ordinary numbers was for me like finding in the background of a familiar photograph a figure I had not seen before and whose presence overthrew the narrative that the photograph had appeared to tell. Prime numbers are a means of escape from the hegemony of the number line. Absent primes, one can go only farther or deeper into the number line, toward infinity in either direction, but the way has no turns. Prime numbers wander. They turn left or right completely. They are a secret society, equipped for more than measuring and commerce, the sorcerers and shamans of the mathematical realm. I am not alone in thinking so. The mathematician Henryk Iwaniec told me, "When you get something unexpected like prime numbers, it seems God given, like something mysterious being found."

It may be beguiling to know that there is no last number, but it is an observation more about infinity than about the hidden qualities of numbers. It might provoke awe, but it is also a mere fact involving the simplest mathematical proof there is, a child can come up with it on his or her own and often does: for any last number n, there is n + 1.

PRIME NUMBERS HAVE their own taxonomy. Twin primes are two apart. Cousin primes are four apart; sexy primes are six apart, six being sex in Latin; and neighbor primes are adjacent at some greater remove. From *Prime Curios!*, by Chris Caldwell and G. L. Honaker, Jr., I know that an absolute prime is prime regardless of how it is arranged: 199, 919, 991. Palindromic primes are the same forward and backward—133020331; they are also called smoothly undulating primes. Tetradic primes are palindromic primes that are also prime backward and when seen in a mirror, such as 11, 101, 1881881.

A beastly prime has 666 in its center. 700666007 is a beastly palindromic prime. A depression prime is a palindromic prime whose interior numbers are the same and smaller than the numbers on the end; 75557, for example. Conversely, plateau primes have interior numbers that are the same and larger than the numbers on the ends, such as 1777771. A circular prime is prime through all its rotations: 1193, 1931, 9311, 3119. There are Cuban primes; Cullen primes; curved digit primes, which have only curved numbers—0, 6, 8, and 9; and straight-digit primes, which have only 1, 4, and 7. A prime from which you can remove numbers and still have a prime is a deletable prime, such as 1937. An emirp is prime even when you reverse its numbers: 389, 983. Invertible primes can be turned upside down and rotated: 109 becomes 601. Gigantic primes have more than 10,000 digits; holey primes have only numbers with holes (0, 4, 6, 8, and 9). There are Mersenne primes; minimal primes; naughty primes, which are made mostly from zeros, naughts; ordinary primes; Pierpont primes; snowball primes, which are prime even if you haven't finished writing all of the number—73939133; titanic primes; Wagstaff primes; Wall-Sun-Sun primes; Wolstenholme primes; Woodall primes; and Yarborough primes, which have neither 0 nor 1.

The only even prime number is 2. Since all other primes are odd, the interval between any two successive primes has to be even, but no one knows a rule to govern this. The largest known prime exceeds by far the estimates of the number of atoms in the universe.

PRIME NUMBERS, WHICH are the subject of number theory, are the origin figures of pure mathematics—mathematics done, that is, without an interest in being useful in any practical way. Applied mathematics begins with the ability to count and measure and is procedural; pure mathematics is imaginative. That the classifications have about them a suggestion of snobbery and side-taking has a lot to do with the British mathematician G. H. Hardy, who sometimes called pure mathematicians "real mathematicians." In "A Mathematician's Apology," Hardy writes, "Is not the position of an ordinary applied mathematician in some ways a little pathetic? If he wants to be useful, he must work in a humdrum way, and he cannot give full play to his fancy even when he wishes to rise to the heights. 'Imaginary' universes are so much more beautiful than this stupidly constructed 'real' one."

The remoteness of pure mathematics is partly why

Hardy's essay is an apology, but he was also pleased at pure mathematics not being helpful in commerce and especially not in war. He defines usefulness as "knowledge which is likely, now or in the comparatively near future, to contribute to the material comfort of mankind, so that mere intellectual satisfaction is irrelevant." He believes of pure mathematics that "the best of it may, like the best literature, continue to cause intense emotional satisfaction to thousands of people after thousands of years."

The aloofness of pure mathematics and its reverence for thinking infused itself into physics. In April of 1969, Robert Wilson, a physicist who had worked on the Manhattan Project and was the director of the Fermi National Accelerator Laboratory, in Illinois, appeared before Congress to request money for building the accelerator, which was called the 200-BeV Synchrotron. The Synchrotron was a proton accelerator that would make it possible to observe subatomic particles, some of which were theoretical.

Wilson was questioned by Senator John Pastore, a Democrat from Rhode Island, who was sympathetic to science and was hoping for arguments he might use to persuade the accelerator's opponents. Senator Pastore asked if the accelerator involved the security of the country.

"No, sir; I do not believe so," Dr. Wilson said.

"Nothing at all?"

"Nothing at all."

"It has no value in that respect?"

"It only has to do with the respect with which we regard one another, the dignity of men, our love of culture," Wilson said. "It has to do with those things. It has nothing to do with the military," for which he added that he was sorry.

Senator Pastore told him not to be sorry.

"I am not, but I cannot in honesty say it has any such application," Wilson said.

Senator Pastore tried another tack. "Is there anything here that projects us in a position of being competitive with the Russians, with regard to this race?" he asked.

Very little, Wilson said. "Otherwise, it has to do with: Are we good painters, good sculptors, great poets? I mean all the things that we really venerate and honor in our country and are patriotic about. In that sense, this new knowledge has all to do with honor and country, but it has nothing to do directly with defending our country, except to help make it worth defending."

Pure mathematics sometimes finds a practical use, but typically after so long that, according to James Newman, the use is thought to be either a coincidence or evidence that mathematics and the world's deeper workings are mystically aligned. Hardy was not entirely correct about its aloofness, though. In *Mathematics Without Apologies*, a response to Hardy, Michael Harris writes that pure mathematics has figured in "radar, electronic computing, cryptography for e-commerce, and image compression, not to mention control of guided missiles, data mining, or options pricing."

Primes, which are secretive, are also essential to keeping secrets. Much of internet commerce and finance depends on a cryptographic system called RSA, which takes the initials of the men who are said to have come up with it, in 1977—Ron Rivest, Adi Shamir, and Leonard Adleman. RSA works by means of a pair of codes called a private key and a public key. A transaction involving a credit card is scrambled using the public key, sent to a bank, then unscrambled using the private key. The public key is the result of two prime numbers multiplied together.

Typically, each prime is more than a thousand digits long. (The largest known prime, which took several thousand computers nearly a month to find, has seventeen million digits.) The public key can be discovered from the private key, but the private key cannot easily be discovered from the public.

Keys are protected by a principle called the trapdoor function. Trapdoor functions are simple to calculate in one direction but hard to calculate in the other without special information, which is the trapdoor. In the case of RSA, the trapdoor involves a field called modulo math. The most familiar form of modulo math is the clockface. Ten hours from four o'clock on the clockface is two o'clock, not fourteen o'clock, except in military time; this is an example of arithmetic modulo twelve. To know the time at any number of hours into the future, you divide the hours by twelve and apply the remainder to the clockface. A version of a clockface can be composed of any number of hours, and this is what happens with RSA. The trapdoor function is a result of knowing the modulo involved in selecting the keys. Simply finding by luck or application the prime numbers involved in the key is useless without the modulo number.

Henry Cohn, a principal researcher at Microsoft, told me that "cryptographic security is based on conjectures that are unproved and seemingly very difficult to resolve. The security of RSA depends on there not being a fast algorithm to factor a number into primes." An algorithm that would factor huge numbers into primes does not appear to exist, "but it's by no means certain," Cohn said.

Primes seem to many mathematicians to be distributed according to a mysterious pattern, which may or may not exist. The more elusive the pattern is, the more many mathematicians are persuaded that it exists. "When you're trying to make a cryptosystem like RSA, you are depending on the fact that numbers have a good deal of structure," Cohn said. He went on to say, however, that "this structure could come back to bite you, by enabling an attacker to defeat the system. Ultimately, what you'd like is a situation that looks beautifully regular and structured to the legitimate users while looking unapproachably random to hackers. This is a subtle balancing act, and lots of cryptosystems have fallen apart over the years when it turned out there was just a little too much hidden structure that could be taken advantage of." This has not yet happened to RSA.

3.

Occasionally a version of pure mathematics put to practical use misleads. This happened with Johannes Kepler, the German astronomer on whose three laws of planetary motion, published between 1609 and 1619, Newton based his law of gravitation, in 1687.

In a portrait of Kepler painted in 1610, when he was forty-nine, he is wearing a ruff collar, and he is looking into the middle distance, as if he were somewhere else in his mind. Maybe he is only impatient with having to sit for the painter, but it makes him seem as if he might have been difficult to talk to, unless you had something to say that interested him. Kepler is a hinge figure. The physicist Wolfgang Pauli, speaking in 1948, describes him as "a spiritual descendant of the Pythagoreans," devoted to finding a harmony among the proportions in nature,

where for him "all beauty lay," and as a signal interme-
diary "between the earlier, magical-symbolical and the
modern, quantitative-mathematical descriptions of nature." Reading *War and Peace* I feel something like astonishment that a single intelligence contained it, and I
feel something similar when I think of the version of the
heavens that Kepler held in his head.

Kepler exemplifies the premodern belief that practical mysteries conceal the divine. According to Pauli, he
thought that the theorems of geometry "have been in
the spirit of God since eternity," and that God was represented by the sphere. "The Father is the center," Kepler
wrote. "The Son is in the outer surface, and the Holy
Ghost is in the equality of the relation between point and
circumferences," which as a model is more concise and
apprehensible than anything I remember being taught in
Sunday school. In the natural world Kepler saw the arrangement as embodied by the sun and the planets. He
felt compelled to discover the mechanics according to
which the planets revolved, since he believed that they
would reveal the sacred design underpinning creation.

Kepler imagined that the planets' orbits accorded
with the forms of the platonic solids. A platonic solid is
a shape in which all the faces are the same and are triangles, squares, or pentagons—that is, they have three,
four, or five sides, which are all equal to one another.
Fantasy games often have dice in the shapes of the platonic solids.

Much of the following material was challenging for
me, although I don't think it's all that complicated. It requires an ability to imagine shapes from more than one
vantage, though—a talent that studying mathematics
made me aware that I don't have—and also, it's a little

dense. It interests me because Kepler's accomplishment is so enormous. The Chinese, the Indians, the Babylonians, the Egyptians, the Greeks, the Arabs, the Jewish rabbis who set up the Jewish calendar all followed movements in the sky, and so did the Mayans, and surely the Africans did, too. The Greeks saw Mercury, Venus, Mars, Jupiter, and Saturn and thought they were stars, but ones that moved differently from, and erratically compared with, the other stars—planets means wanderers in Greek. Kepler was the first to describe the mathematical laws governing celestial mechanics—the paths the planets followed and the pace at which they followed them—to find an order, that is, in something that had seemed disorderly. Carl Sagan called Kepler "the first astrophysicist and the last scientific astrologer." If chess were played with figures from mathematical history as pieces, Kepler might be one of the pieces.

So here goes: At each vertex of a platonic solid—that is, corner where they form an angle—the same number of faces meet. A cube is a platonic solid. There are four others: the tetrahedron, the octahedron, the icosahedron, and the dodecahedron. The tetrahedron is a pyramid with three equilateral triangles meeting at each vertex. The octahedron has eight faces and four equilateral triangles meeting at each vertex. An octahedron looks like a pyramid sitting on top of an upside-down pyramid. An icosahedron has five triangles meeting at each vertex, and I can't think of anything it looks like. The dodecahedron has three pentagons at each vertex and twenty sides, forming a pattern like the pattern on the side of a soccer ball. Euclid proved that there can be only five platonic solids, because six equilateral triangles arranged around a vertex form a circle, being 360 degrees, and therefore lie flat.

Kepler's belief in a correspondence between the solar system and an arrangement of the platonic solids descended from Plato, who says in *Timaeus* that the world is made from earth, air, water, and fire, and that each of these is formed from one of the solids. Earth, being "the most immoveable," is formed from the cube, which has the most stable base. Fire, which dissolves other elements by its sharpness, is made of the tetrahedron, which is pointy. The octahedron is assigned to air for being something like smooth. Water is made from the icosahedron, which is complex and heavy and able to crush fire and earth. The dodecahedron represents the universe itself, since it most closely approaches a sphere.

Kepler believed that each planet, according to its orbit, could be placed inside a solid that is enclosed by the next larger orbit and, like a nesting doll, is itself inside another solid. By their distances from the sun, from least to greatest, Mercury sits in an octahedron enclosed by Venus; Venus sits in an icosahedron enclosed by Earth; Earth sits inside a dodecahedron enclosed by Mars; Mars sits inside a tetrahedron enclosed by Jupiter; and Jupiter sits inside a cube enclosed by Saturn. The orbit of the inner planet is tangent to the center of the face of the solid enclosing it, meaning that it touches the face at a single point, and the orbit of the outer planet travels through the solid's vertices. The arrangement requires that the planets' orbits be circles, also a platonic notion. For Kepler, the planets moving in a circle demonstrated a divine pattern.

Eventually, Kepler noticed that if he drew a line from the sun to a planet and a similar line at an interval in the planet's orbit, either of hours or days, the area enclosed is always the same no matter where the planet is, so long as the interval of time is the same. Such a circumstance could happen only if the planet travels faster when it is closer to the sun than it does when it's farther away. And this is possible only if the orbit is an ellipse, which is not a divine form.

Kepler's three laws of planetary motion defined the correct orbits. The first law is that the orbit of a planet is essentially an ellipse. The second is that the progress of a line drawn between a planet and the sun encloses equal areas during equal intervals of time. And the third is that the square of the period of the orbit is proportional to the cube of the semi-major axis of its orbit. According to Amie, in whose field Kepler's work figures, this means that if you square the planet's orbital year and divide it by the cube of the distance to the sun, the number is the same for all planets. Finis.

4.

The more I learned about the habits of mathematicians, the more it seemed that they approached their work in much the same way that writers and artists approach theirs, which didn't make doing math easier, but it interested me. So far as I understand it, mathematical creativity involves the same stages that creative thought involves in any discipline or art. Like novelists and musicians, mathematicians produce thought objects that have no presence in the physical world. (Anna Karenina is no more actual than a thought about Anna Karenina.) Like other artists, mathematicians also have the run of a world that others hardly or only rarely visit or don't travel very far into. For mathematicians, though, this territory has more rules than it does for others. Also, what is different for mathematicians is that all of them agree about the contents of that world, so far as they are acquainted with it, and all of them see the same objects within it, even though the objects are notional. No one's version, so long as it is accurate, is more correct than someone else's. Parts of this world are densely inhabited and parts are hardly settled. Parts have been visited by only a few people, and parts are unknown like the dark places on a medieval map; somewhere among this territory would be where the proof of the Collatz conjecture resides. The known parts are ephemeral but also concrete for being true and more reliable and everlasting than any object in the physical world. Two people who do not share a language or understand a word the other is saying can do mathematics with each other, silently, like a meditation.

An imaginary world's being infallible is very strange. This spectral quality is bewildering even to mathematicians. In text accompanying his portrait in *Mathematicians: An Outer View of the Inner World*, a series of portraits by the photographer Mariana Cook, the mathematician John Conway says, "It's quite astonishing, and I still don't understand it, despite having been a mathematician all my life. How can things be there without actually being there?"

5.

We don't often encounter the limits of our intelligence, but the way I struggled with algebra sometimes made me wonder if I was finding my own. At such times I felt myself to be a poorly equipped version of human possibility, sort of a discard. I was also almost daily reminded of how some things needed to be learned more slowly. Meanwhile, I was harassed by my upbringing to believe that I had to work quickly; any half-smart person could work out a problem given sufficient time. I found these attitudes difficult to combine. Sometimes I realized that I was talking to different parts of myself, and the exchange was not polite.

Occasionally I got good at operations that were hard at first. This happened with factoring, a process in algebra of simplifying expressions and with expanding, which is the opposite of factoring. The axioms of arithmetic imply that when you expand $(a + b)^2$, for example, you get $a^2 + 2ab + b^2$ in the following way: $(a + b)^2$ is equal to $(a + b)(a + b)$. Each term in one parentheses multiplies the terms in the other: $a \times a = a^2$; $a \times b = ab$; $b \times a = ab$; $b \times b = b^2$. Combining the terms, $a^2 + ab + ab + b^2 = a^2 + 2ab + b^2$. In a similar way, $a^2 - b^2$, a squared number subtracted from another squared number, called a difference of squares, becomes $(a - b)(a + b)$, which becomes $a^2 + ab - ab - b^2$, which is $a^2 - b^2$.

Simple, but I really liked it. As the formulas became more complicated, there were more steps, each of which followed from the one before it, so that in addition to finding the answer, there was the pleasure of enacting a procedure properly, plus no textbook skipped the steps. Each time I turned a page and saw more factoring, I was pleased. It was like being good at spelling and wanting to be asked more words. Accompanying my pleasure, though, was a voice saying, "Listen, Slick, this is algebra for kids. We can throw problems at you, you won't even know what they *mean*."

Sometimes I dreamed that there were numbers falling from the sky into chasms I couldn't see the bottom of.

6.

Aside from learning starter mathematics, I became interested in trying to understand what mathematics is. Variously, I thought, a structure, a terrain, a republic, a façade concealing an infinite kingdom, an archive that goes on forever, a library of manuscripts and artifacts penned behind a series of more and more elaborate locks and doors.

For as long as I can remember I have been drawn to the hidden life—scenes that play behind the eyes before sleep, the fleeting impressions from the periphery of one's vision that are there and not there, reveries, anything that resides on the border of consciousness. I like it in all its guises and representations and almost anywhere it shows up, but I am especially drawn to objects and places that are underground. The bottom of the ocean or a pond or the beds of underground rivers.

I often dream that I am in a basement, or descending stairs to the subway, or walking in a cave, or swimming toward lights at the bottom of a river, or finding rooms belowground that no one else knows about and that aren't there when I try to go back to them. The world of symbols is fairly impenetrable to me, but I am not so thick that I don't see that the underground stands for the Unconscious. I sometimes think of the Unconscious as a series of rooms, each opening out from the next. In one of them, perhaps, a book lies open on a table. In another an old woman sits in a rocking chair while rain beats against a window. On a wall in another is writing that you can almost read, or a mural depicting a scene that turns up later in a dream or perhaps as a premonition. Or maybe not as a series of rooms but as a landscape. It has weather and there is night and day, but it is not always a landscape I recognize, and it changes all the time, as if each vista were a fragment of another one, like the planes in a Cubist painting. I am drawn to the Unconscious for the reasons I assume most people are, which is the belief that something it contains has the power to release you from torment. Or that something lost can be recovered there. Or that it is rich in inspiration and productive of inexplicable intimations and feelings.

Occasionally I hear about subterranean places in New York City, and I go visit them, such as the corridors and tunnels under Grand Central or the railroad lines along the West Side, by the Hudson River, where until a few years ago, when the railroad police began chasing them out, squatters lived in the cinder-block chambers that the railroad had built in the walls beside the tracks for its workers to use while they were constructing the tunnels. The railroad had lit them but they weren't lit anymore, and the darkness was so complete that the men and women who inhabited them couldn't see their hands in front of their faces.

I had a hobby of looking for tunnels that I had heard were dug by bootleggers between their basements and the rivers; tunnels in Chinatown built during the Tong Wars for fighters to disappear into; and a section of the aqueduct in the Bronx that someone told me went all the way to Manhattan. One night I persuaded a friend who had bolt cutters to cut the lock on a chain that held a gated door closed over a tunnel in Central Park on one of the transverses. The next night I came back and went through the door. There was a long vaulted tunnel of stone with a dirt floor leading to the basement of the pump house at the southern end of the Reservoir, which I explored with a flashlight for as long as I thought was safe. Not long after that I took a long subway and bus ride to the end of Brooklyn, where it plays out into marshland and sand dunes, and walked up a hill past gates that no longer worked to visit the shoebox-shaped concrete holes where surface-to-air missiles were kept during the 1950s and '60s. It was so far away from the rest of the city and so quiet that I might as well have been in the desert listening to the wind whistle.

The more I read about mathematics, the more it seemed to me that unconscious thought was responsible for a lot of it. The French mathematician Jacques Hadamard published *The Psychology of Invention in the Mathematical Field* in 1945, having sent a questionnaire to a number of mathematicians and physicists asking about their methods. Hadamard had been inspired by ideas he heard described by Poincaré in a talk on mathematical invention that Poincaré had given in Paris in 1908.

Hadamard describes four stages of mathematical creation: preparation, incubation, illumination, and verification. Preparation involves the gestures made on a first encounter, a throwing of what one knows at the problem. (The artist Saul Steinberg once told me that he began each working day by setting himself a problem.) For some mathematicians the first encounter is aggressive, a frontal assault, and for others, such as the French mathematician Alain Connes, it is oblique, because Connes believes that too assertive an approach can leave a mathematician not having solved the problem but having run out of tactics.

Amie said that one of her teachers had told her, "A good problem fights back." Beyond the obvious inference, I took this to mean that such a problem might welcome attention but diffidently. Or that it is something like hills in the distance observed through a mist. The hills are apparent but the way there is not, or even if there is a way there. An approach that appears promising might collapse immediately, but it might also fall apart after a lot of work.

Incubation describes that period when, having been rebuffed, the mathematician might put the problem aside, hoping or perhaps expecting that some other part of his or her mind would continue to work on it, a method that Einstein told Hadamard he favored. Making associations among disparate means of progress takes time. Connes says that sometimes he finds it helpful to work in a field parallel to the one occupied by his problem, one that is perhaps a little easier and has more opportunities for engagement, in which he might develop a framework for his thoughts and approach the parallel problem indirectly. Thinking in a way that isn't linear, that is abstracted and involves a dissociated state of mind, is characteristic of what the neuroscientist Jean-Pierre Changeux describes as "a blackout of rational thought." He says this in *Conversations on Mind, Matter, and Mathematics*, which is a collection of exchanges between him and Alain Connes and a book I liked so much that I read it several times.

Illumination ends incubation. Maybe the mathematician drank too much coffee, or drank coffee and isn't used to it, something that happened to Poincaré. Maybe he or she worked to the point of exhaustion and the defenses that might be invoked against psychic intrusions are fatigued. Maybe a long walk stimulates the imagination. Or waking from a nap.

Regardless, something previously unseen seems suddenly at hand. This is a sanctified state of being. Connes says of such times, "I couldn't keep the tears from coming to my eyes." All that remains, Hadamard says, is verifying one's vision while hoping that it is genuine and doesn't collapse.

7.

Mathematics is not a practice that one ought necessarily to expect to be better at as one gets older, but I did expect, somehow, to be better at it. I had the feeling that the limitations I had placed on myself when I was a boy were matters I had disposed of, partly by showing myself that I wasn't inept in the adult world. I also felt that I was returning to the encounter with more tools. Many of the subversive processes that overtook me in attempting math a second time, however, were bound into my being, my nature; they *were* me, at least a part of me, those hesitations and reluctances and fears of incapacity and, although it pains me a bit to say this, the embarrassment I had felt at being found to be insufficient. This was a struggle I hadn't anticipated.

Hegel somewhere remarks that the reason there are so many examples of childhood prodigies in mathematics is that mathematics doesn't involve (as music doesn't either) a grasp of the complexities of mature life, which need experience to make sense of them. It is a common-place in mathematics that young mathematicians do more consequential work than older ones. This seems to be a judgment that mathematics can't escape. I have heard a number of explanations for this, aside from the obvious one that vitality and exuberance matter. Experience can make you hesitant is one. Also, children can see patterns at an abstract level that are obscured for older people, who see the patterns embedded in a larger structure and likely a familiar one that they aren't able to see anew. Older mathematicians have independently to learn complicated new mathematics that younger mathematicians are fluent in, from having studied them recently in school. Also, young mathematicians don't know what to be afraid of, and so can make breakthroughs on problems that older mathematicians have given up on. Andrew Granville, a mathematician born in 1962, once told an interviewer, "To crush a great problem, it often means you have no respect for where you are told not to go." And, finally, the absence of a collection of elders who can impede the progress of a young mathematician whose work doesn't fit their tastes, since mathematics is largely an individual task and taste does not figure extravagantly in mathematics, although, as elsewhere, favoritism of gender and race sometimes figures in academic life.

Serious mathematics requires a single-mindedness that Hadamard defines as "a tenacious continuity of attention," and as "a voluntary faithfulness to an idea," phrases that reverberate poetically. A mathematician might spend a career on only a few problems. Georges-Louis Leclerc, a French mathematician of the eighteenth century, thought that genius was a capacity for great patience. Henryk Iwaniec, who was born in Poland in 1947, told me that persistence was also necessary for a mathematician and that arriving at a result was a matter of "really conquering the turf to the end." Mathematics for Iwaniec, as for many mathematicians, is an intimate endeavor. "We are driven by emotions," he said. "Mathematics is still a beautiful personal challenge."

HADAMARD DIVIDES A mathematician's methods into two categories. One consists of the time spent with a problem, beginning with selecting it. The second involves the work that goes on unconsciously. From this engagement an answer emerges, sometimes entire or else in the form of an organizing principle. Hadamard quotes a nineteenth-century mental calculator named Ferrol who writes to the German mathematician August Möbius, "It often seems to me, especially when I am alone, that I find myself in another world. Ideas of numbers seem to live. Suddenly, questions of any kind rise before my eyes with their answers."

Furthermore, Hadamard says, there are two types of invention. One type sees a result and imagines the means of achieving it. The other knows an answer and pursues ways of using it. One type of mathematician sees things at close hand, and the other sees them from a remove that is congenial to making generalizations between or among theorems or fields.

Breakthroughs on big problems are exhilarating. In *Mathematics Without Apologies*, Michael Harris quotes Marcus du Sautoy, saying "Once you have experienced a buzz of cracking an unsolved problem or discovering a new mathematical concept, you spend your life trying to repeat that feeling." For the French mathematician

Marie-France Vignéras, also quoted by Harris, "it happens suddenly: one direction becomes more dense, or more luminous. To experience this intense moment is the reason why I became a mathematician." For Bourbaki, the pseudonym of a group of French mathematicians from the middle part of the twentieth century, a form of intuition, not "popular sense-intuition, but rather a kind of direct divination," sheds light "at one stroke in an unexpected direction," and this "illumines with a new light the mathematical landscape."

8.

Mathematical investigations depend on the tools at hand, but mathematicians can also invent new tools or find new uses for established ones. As with all arts, the more one can define, or perhaps restrict, one's intentions, the more progress is likely, although many of the problems in deep mathematics are so difficult to approach that progress is sometimes made only incrementally, over long periods of time, and often by people not necessarily working with each other or even in the same era. A mathematician at work on a deep problem might have only companions from history or, conversely, ones from his or her posterity.

A pure mathematician seeking a result is pursuing something that he or she believes exists. The work isn't being done in the expectation of finding a surprising result that hadn't been contemplated, although possibly this happens. Pure mathematics is not generally haphazard, though. A mathematician typically knows what he or she is pursuing—the steps necessary to approach a big problem make a mathematician more liable to think, Something might happen, than, Anything can happen.

The mathematician Gregory Chaitin once told an interviewer, though, "Science is the same idea as magic: that there are hidden things behind everyday appearances. Everyday appearance is not the real reality."

Spring

1.

As I progressed, my eye progressed, and more than solving algebra problems by grasping their design, I became more clever at reading questions. To do better, though, I had to become vigilant. For someone who thought that there were shortcuts and faster passageways to learning, this was unwelcome. I had never understood that learning needs to be done patiently. One can be impatient to learn or for learning in general, but that is a matter of temperament. I am having to learn how to learn. In school they expect you to learn, but they don't teach you how to learn, at least they didn't in my childhood.

I am accustomed to remembering what I hear and being able to draw on it. Learning algebra requires a secondary use of information, though, a sorting and referencing, a repetition of experience, so that it actually is experience. With algebra I'm not simply collecting information, I am having to classify and comprehend it. We do this naturally as children in classrooms, partly because the distant future seems as if it will never arrive, but it is a different matter to be older and feel that one's capital of time is remorselessly diminishing. Such a consideration adds a complicating haste and impatience.

. . .

DIVIDING THE FRACTION 7/2 by 2, I confused the properties of exponents, and thought that the product is 7,
since $7 \times 2 = 14$ and $14/2 = 7$, when in fact the answer is
7/4, since dividing a fraction by 2 is the same as multiplying by 1/2, but I got the answer wrong and got angry at
math and called Amie, and she wouldn't talk to me until
I calmed down. She wasn't always calm, either. Once I
heard Benson, her husband, in the background say, "Why
are you yelling at him?"

When I had worn Amie's patience too thin, I would
call Deane Yang, my friend who is a mathematics professor at NYU. "The way you remember procedures is you
remember why," he said.

"Because?"

"Because people learn math as a collection of procedures," he said. "When things get difficult, they're lost,
and math becomes religion class. The teacher says what's
right and wrong, and for all you know math came out of
the sky, and some prophet told you how to do it, and it's
just blind belief then. The goal is to take on questions that
appear to be complicated, and to recognize that a complicated question can be broken down into simpler questions, some of which can be answered independently of
each other."

Among the algebra shortcuts I have been taking is
writing out only the portion of an equation that I am
working on, while keeping the rest in my head, which is
an efficient way to lose control of plus and minus signs.
$(x + 7) - (y - 4)$, for example, is an expression. (An equation
has an equal sign: $(x + 7) - (y - 4) = 21$.) As for plus and
minus signs, the minus between the two entries means

that the y is actually –y and –4 is 4, since the minus sign is the equivalent of a –1 in front of the parentheses and –1 × y = –y and –1 × –4 = 4, since a negative number times a negative number is a positive number. In the midst of the calculations, the signs change, and when the formulas are longer, and there are more steps, I tend to get lost.

"With math you have to be very, very disciplined," Deane said. "Normally with algebra, you're trying to make something complicated simpler, but often, temporarily, you have to make it more complicated. The only way to be properly disciplined is to remember exactly what you're allowed to do, and what you're not allowed to do. You have to write everything down, line by line. Math is very painstaking."

These remarks had an almost Zen-like forcefulness for me. They were both abstract and practical, they spoke to my distress, and for a while things got better.

2.

To respond to Gregory Chaitin, the case that there is a reality different from the apparent one can be presented in an orderly way, beginning with the assertion that the reality we perceive is not objective reality, it only appears so. Hummingbirds and certain other animals see more colors than we see. To them the world looks more as it is. There are other hidden things behind all the things that we can see, even if all we allow is scale. Our bodies as a pattern of atoms or Saturn shown at its actual size are images that are only conceptually possible.

In *We*, a novel published in 1921 by the Russian writer

Yevgeny Zamyatin, a practical case for a separate realm is made. "We have never seen any curve or solid corresponding to my square root of minus one," Zamyatin writes. "The horrifying part of the situation is that there *exist* such curves or solids. Unseen by us they do exist, they must, inevitably; for in mathematics, as on a screen, strange, sharp shadows appear before us. One must remember that mathematics, like death, never makes mistakes. If we are unable to see those irrational curves or solids, it means only that they inevitably possess a whole immense world somewhere beneath the surface of our life."

To believe in ideas is to believe in the possibility of a transcendent realm. Perhaps in a practical transcendent, where ideas reside, where the material of dreams comes from, something less like Plato's non-spatiotemporal realm than Jung's collective unconscious, which is patterned on Plato's example but exemplifies humanist principles and not strictly metaphysical ones. Ideas do not seem to be part of physical reality, unless one believes that they exist only as transactions within our brains, which, of course, is possible, too.

Mathematicians who aren't comfortable with Platonism are nevertheless comfortable with there being an exalted feeling to doing mathematics. Awe and wonder are the feelings Amie says she has doing math. Uniqueness and passion are what the Harvard professor Barry Mazur says he experiences. So where are they located imaginally when they have these feelings? Is the explanation simply self-enlarging; are they simply smarter than others who haven't thought of such things, or maybe can't think of such things; is that all there is to it? Or is it, by means of endeavor and talent and aptitude, a matter of being enfolded within, having access to, a territory not always close at hand and not easy to reach?

With the ancients, admittance to this territory depended on a sponsor, a muse. It required preparations, grace, and a plea's being made. There is still a superstitious feeling among creative people about being admitted to and allowed to remain in a realm that appears to be separate from ordinary life.

"It's still possible, you close your eyes, you forget about all your problems, and you try with your intuition to reach the Platonic world of ideas," Chaitin told Robert Lawrence Kuhn in Kuhn's broadcast series *Closer to Truth*.

Of course, these notions are speculations. The mathematician Michael Harris defines "speculative philosophy" as "taking phenomena as symptomatic of something that remains concealed," a remark I admire for its concision.

PLATONISM: THE FIRST philosophical position a mathematician encounters is whether mathematics is created or discovered. It is practically unavoidable. If I think that mathematics is discovered, then it has a perfection, an orderliness, and a permanence that a human mind cannot achieve on its own. If I think mathematics is created, then it is necessarily imperfect, opportunistically ordered, and parts of it are impermanent, since it reflects human misconceptions and shortcomings, which require revision, and perfection is an abstraction anyway.

Mathematicians don't tend to endorse Platonism, if they do, because they have read Plato. Platonism in mathematics is not something that Plato thought about, except obliquely. It is a modern more than an ancient concern. Plato's name was attached to mystical assertions about mathematics in 1934 by the Swiss mathematician Paul Bernays at a talk in Geneva. Bernays noted that both Euclid and the German mathematician David Hilbert discussed mathematical objects as "cut off from all links with the reflecting subject," by which he meant human beings. "Since this tendency asserted itself especially in the philosophy of Plato, allow me to call it 'Platonism,'" Bernays said, invoking the non-spatiotemporal realm, the forms of which were replicated in lesser versions by the Demiurge whom Plato believed had fabricated the world.

Among Western theologians of the Middle Ages, mathematics was generally regarded as one of the things that God created, either from the chaos of the void, which resembled the non-spatiotemporal realm, or, as Christians began to think in the second century CE under the influence of Philo, from nothing, ex nihilo. Or they thought of mathematics sometimes as a tool that God made available so that his works might be understood if only partially. Alain Connes says that Blaise Pascal, whom Connes describes as "the most famous mathematical mystic," thought of the infinitely small and the infinitely large as mysteries that nature had posed so that they might be admired, even while they could not be understood.

In *Timaeus* Plato writes that God lit the heavens so that the animals, of which humans were the highest form, "might participate in number, learning arithmetic from the revolutions" of the stars and the planets. So that these could be observed in their entirety, he created day and night. The sky provided a means of transit from the mathematics of visible things to the mathematics of invisible things.

A Platonist believes that the native territory of numbers and mathematics is somewhere else than in physical reality. This elsewhere is abstract, indifferent to us, was here before us, and will be after. It is neither eternal nor everlasting, since those are temporal distinctions and the non-spatiotemporal realm is atemporal; it simply is. (Jung's collective unconscious is spatiotemporal, requiring the presence of human beings as something like channelers or vessels.) This adjacent reality is not connected in any clear way to the physical world, but it is somehow accessible to human thinking.

THE EXPLANATIONS FOR how mathematics is apprehended and how mathematicians might be in touch with a non-spatiotemporal realm include mathematics being viewed as something like a dream—a serial vision or a narrative, that is—apparently fashioned within the Unconscious, which makes of the Unconscious a kind of worldly conduit to the non-spatiotemporal realm. Mathematics then is an activity that the Unconscious performs and delivers to waking life, which is how Poincaré and Hadamard saw it. This exchange is one-sided in that the Unconscious can produce mathematics but a mathematician cannot commission work; he or she can only inquire into whether the Unconscious has any results to offer. Pleading and attentiveness are the only means of engaging it.

The influential German mathematician and logician Kurt Gödel was a Platonist. In the paper "On Gödel's 'Platonism,'" Pierre Cassou-Noguès quotes passages from Gödel's papers saying that "something in us different from our ego" provides results, and that the question in mathematics is "to find out what we have perhaps unconsciously created." Gödel writes that mathematical ideas "form an objective reality of their own, which we cannot create or change, but only perceive and describe." And,

"Mathematics describes a non-sensual reality, which exists independently both of the acts and of the dispositions of the human mind and is only perceived, and probably perceived very incompletely, by the human mind." According to Cassou-Noguès, among Gödel's papers there is a note from 1946 in which "Gödel mentions the possibility that we see mathematical truths in God's mind." This might make a superior mathematician a type of clairvoyant.

"I believe that mathematical reality lies outside us, that our function is to discover or *observe* it, and that the theorems which we prove, and which we describe grandiloquently as our 'creations,' are simply our notes of our observations," Hardy writes in *A Mathematician's Apology*.

It seems likely that wherever mathematical reality resides, it does not appear to be entirely within us, and that we do not bring it into being simply by thinking of it, since if that were the case, different people might think of different mathematics, which might happen if there were no general agreement of the need for proof, and there wouldn't be, since there would not likely be any general agreements at all. There would be only opinions and convictions about what is correct mathematics. Like a faith then, mathematics could split into sects. Also, it would seem that if math were created instead of discovered then the rules could change by consent, as in a game, which they can't. (Deciding that 1 is not prime isn't changing a rule, it is coming to an agreement about a case; deciding that $3 + 3 = 7$ would be changing a rule.)

IF THE THINGS that we take in with our senses are not everything that there is to reality, then where, if anywhere, is the rest of it? A neurological explanation might say that the missing parts are in our minds, in consciousness, somehow. It doesn't ask much of our reasoning to believe, as Carl Jung did, that human beings are born with a primitive cultural knowledge, a heritage of symbols and archetypes, and that some part of this inventory is productive of pure mathematics or at least sympathetic to producing it.

According to Plato, we are fitted from birth to recognize mathematical truths, specifically geometrical ones. The capacity is part of some collective human reservoir. In *Meno*, he has Socrates, by means of questioning, lead an uneducated boy to discover truths about the areas of squares that Socrates draws in the sand. Since it appears plain that the boy could not have learned these truths, Socrates concludes that he must be recollecting something he knew from before he was born. This demonstrates that there is such a thing as true knowledge, knowledge of the eternal, and that the "soul has been forever in a state of knowledge," Socrates says.

Some parts of mathematical knowledge do seem inherent and intuitive. Mathematicians like to point out the complicated mathematics involved in crossing a crowded street, which we solve without being aware that we are solving it. This seems to suggest either that we know certain mathematical truths without having learned them or that we learn them without knowing we are learning them.

3.

Once I encountered Platonism, I was a goner. Mathematics mostly rebuffed me, since I could perform it usually only badly, but *thinking* about mathematics I could do, because anyone can. The question of whether mathematics is created or discovered appeals to me deeply. I get a little avid when I talk about it, and I am brought up short each time I realize that someone I'm speaking to isn't as keen about it as I am.

It turns out there are degrees of Platonism, if you can stand it. An Absolute Platonism regards mathematics as perfect and timeless and independent of human beings. A Strict Platonism believes that a complete version of mathematics exists, has been being revealed for thousands of years, and may be inexhaustible. A slightly less emphatic version asserts that mathematics is a description of an adjacent world, a type of inventory or a census. The mathematician who said that mathematics is "a proto-text whose existence is only postulated" was expressing a Platonic point of view.

About 65 percent of mathematicians are Platonists, at least that is the figure I find most often. The bulk of them tend to endorse a type of passive Platonism that allows that mathematics has aspects that human minds cannot account for, its capaciousness, for one thing, as well as objects that are inherently too large and complex for human minds to encompass.

For the mathematician and physicist Roger Penrose, to say that a mathematical statement is Platonic is to assert that it is true objectively—true, that is, whether anyone believes it or not. The assessment becomes a declaration of authenticity. Instead of believing that mathematics resides in a non-spatiotemporal realm, Penrose thinks of mathematics as encompassing a realm of its own. He gives the example of a Mandelbrot set. A Mandelbrot set is produced by mapping points involving complex numbers.

A complex number is a number that can be written as a +
b*i*, in which a and b are real numbers and $i = \sqrt{-1}$, a number designed to have the property $i^2 = -1$. Since no real
number fulfills this equation, any number that includes
i is called an imaginary number. When drawn, Mandelbrot sets are made of star shapes and waves and spikes and
spirals and clusters of other geometric forms that repeat
ceaselessly in finer and finer detail; some of the versions
I have watched in videos look a little like perpetual and
ever-diminishing depictions of the Hokusai woodblock
The Great Wave off Konagawa. Mandelbrot sets do not exist
"within our minds, for no one can fully comprehend the
set's endless variety and unlimited complication," Penrose
writes. Instead they can reside only "within the Platonic
world of mathematical forms." This also shades toward a
mystical interpretation.

The German mathematician and logician Gottlob
Frege also saw mathematics as having a truthfulness of an
ideal kind, meaning that a mathematical statement is true
"whether anyone takes it to be true. It needs no bearer. It
is not true for the first time when it is discovered."

Mathematics in this case is like a river, which exists
before it is given a name.

PARTLY IT'S OBVIOUS that mathematics exists without us,
and partly it isn't obvious. Obvious because if we vanish
and leave no trace, painting and sculpture as we practice
them might be lost, writing might be lost, music might be
lost, but mathematics will exist whether we do it or not.
Numbers won't arrange themselves differently or work
in different ways or have different properties, because
they haven't been used for a while or are being used by

different creatures. Two objects added to another object will be three objects. The numbers 2 and 3 will still have the qualities of being prime, although those qualities might be latent until a new species identifies them. A replacement species might give numbers a different name, but the numbers themselves, with their attributes, are inviolable. According to Warren S. McCulloch in his 1960 lecture "What Is a Number, That a Man May Know It, and a Man, That He May Know a Number?," around 500 CE St. Augustine wrote, "7 and 3 are 10; 7 and 3 have always been 10; 7 and 3 at no time and in no way have ever been anything but 10; 7 and 3 will always be 10. I said that these indestructible truths of arithmetic are common to all who reason." Augustine regarded each of his remarks as an eternal verity, "true regardless of the time and place of its utterance," McCulloch writes. "Each he calls an idea in the Mind of God, which we can understand but can never comprehend."

Not obvious has factions, too. There is a belief that numbers represent actual objects and another that they represent mental ones. Plato thought that numbers were abstract objects, and Aristotle thought that they were abstractions from actual objects. In *Platonism and Anti-Platonism in Mathematics*, Mark Balaguer says that the thinking in favor of actual objects would say that when 2 objects and 1 object are added, we have 3 objects, no matter what objects they are or whether or not they are identified. As for mental objects, 2 objects added to another is a remark about what 1, 2, and 3 stand for and says nothing about any objects involved.

The main arguments opposed to Platonism are Formalism, Constructivism, and Intuitionism. Formalism is the brainiest. It says that mathematics is simply a pastime, a game played by rules and properties that have meaning only within the game. In Formalism, 4, say, is used to play mathematics the way a pawn is used to play chess. Axioms, theorems, numbers, proofs, symbols, and equations, the whole paraphernalia of mathematics, are immaterial in the sense that they have no existence except as they appear on the page. They are gestures signifying intent and pursuit, and while seeming to be the most literal of statements and intensely specific they are also totally conceptual. Equations are symbols arranged serially. Formalism seems akin to the neurological notion that dreams are images that are too insignificant to find their way into waking thought and are being shed during sleep. According to this reasoning, to suggest that the images have deeper meaning is to misunderstand them fundamentally. It makes fabulists of figures such as Freud and Jung.

Formalism appeared during the late nineteenth and early twentieth centuries as a means of making mathematics more rigorous by revising its principles so as to rid them of intuitive assumptions, intuition being then regarded as a weakening force in mathematics. Formalism is identified with the mathematician David Hilbert. It has a surrealist quality. In relation to Euclid, Hilbert says, "One must be able to say at all times—instead of points, straight lines, and planes—tables, chairs, and beer mugs." It seems fanciful and strange, the notion of a satirist dismissing the imagination as a source of insight and instruction, but it is also an attempt to clarify mathematics and to make it more infallible, more unshakable than it already is. Whatever else Formalism accomplishes, though, it doesn't give mathematics a closer correspondence to reality, since to a Formalist there is no correspondence to begin with.

Formalism excludes all but the surface of things.

Creative thought can perhaps be carried out according to rules, especially rules that helpfully limit an apparent infinity of choices, but the game wouldn't seem able to advance without intuitions about how to use the rules in new ways. A divorce from the realm of psychic associations may not even be possible, though, considering all that the psyche appears to harbor and give rise to.

4.

An amateur dissenter from Formalism might wonder timidly if it doesn't allow even the origins of Formalism to be Platonic, since the non-spatiotemporal realm includes concepts and ideas in their perfect form. Formalism says only that humans take part in mathematics by agreeing to certain arrangements. Endorsing Formalism would seem equivalent to saying that any thought is essentially a succinct neurological event, a synapse turning on and then off. A thought shared among a number of people says nothing about the character or truthfulness of the thought, it says only that different human brains can have similar thoughts or conduct similar operations. Formalism appears to see human beings and their thoughts as being at the center of existence, whereas Platonism sees them as existing at some remove.

Formalism and Platonism are also twinned, but retroactively. Or perhaps, like many themes in mathematics, they are parallel. A Formalist would say that numbers exist to play the game, and that the rules reflect their attributes. The numbers 2 and 3 and 5 and 7 are prime because the quality of being prime is among the rules, but this doesn't really respond to the possibility of primeness being an attribute independent of the game, or that the game is played with entities that existed before anyone thought to use them in a game.

INTUITIONISM ALSO BELIEVES that mathematics is the mind's creation, and that it doesn't describe anything realistic. A more emphatic version of Intuitionism is Constructivism, which believes that the mind constructs mathematics according to intuitive dispositions such as the inclination to assess the world in terms of numbers. The basic principles of arithmetic emerge from a human capacity for counting and for handling simple arithmetic procedures. The mathematics that we have, in other words, is the mathematics that we have brought about and can do. To a Constructivist, a mathematical formula is a tool or a prop, and the mind is something like a theater in which mathematics performs.

The belief that there are no such things as abstract objects, and that therefore mathematical statements, which rely on abstract objects, are not true, is called Fictionalism. They aren't false, either, so much as fanciful. According to an argument I find in the Stanford Encyclopedia of Philosophy under *Formalism*, to say that 3 has the quality of being prime is like saying that among the tooth fairy's qualities is generosity.

THE SIMPLEST, OLDEST, and most obdurate argument dogging Platonism is the epistemological argument. It says that if mathematical objects are abstract, then they exist outside of space-time, and human beings, who live in space-time,

could not know about them. Since human beings know
about them, they can't be abstract, so Platonism is false.

I might offer that there is more than one explanation of
how a person could be connected to a non-spatiotemporal
realm, all of them conjectural, of course. There is Plato's
belief in *Meno* that our souls know about abstract objects
such as the forms and that learning mathematics is a matter of remembering what we know. Furthermore, one
can infer from things that Socrates says in Books VI and
VII of *Republic* that Plato believes that a part of the soul
itself is non-spatiotemporal or has the potential to be. I
am partial to St. Augustine in *On the Free Choice of the Will*,
who writes, "The double of any number is found to be exactly as far from that number as it is from the beginning
of the series," which for the purposes of simplicity might
be thought of as zero. "How do we find this changeless,
firm and unbroken rule persisting throughout the numerical series? How can any phantasy or phantasm yield such
certain truth about numbers which are innumerable? We
must know this by the inner light, of which bodily sense
knows nothing."

Mathematics, in its independence and timelessness, becomes also a type of self-intoxication, and self-justifying,
a creator's romance with his or her creation, qualities that
more than one culture have attributed to God.

5.

A citizen Platonist responds (briefly) to further dissent
before turning out the lights: In *The World of Mathematics*,
James Newman says that the discovery of non-Euclidean

geometry, in the early nineteenth century, meant the end of metaphysical speculations about the origins of mathematics, since non-Euclidean geometry involves figures such as spheres and other objects with properties that are different from those of triangles and squares and circles on a flat plane and also includes figures that can be extended to spaces of three and more dimensions, that such a field is a creation of the human mind. Wouldn't it also be true, however, to say that, as with the principles of geometry, those of non-Euclidean geometry were latent and waiting to be discovered? As with Euclidean geometry, mathematicians didn't invent the forms of non-Euclidean geometry, they observed and described them. The forms being identified by means of intellectual effort does not address their origins, since, like all processes of thought, mathematics is the result of intellectual effort.

For Timothy Gowers, in *Mathematics: A Very Short Introduction*, the case of i appears to refute Platonism: i requires that a person think abstractly, Gowers says, because either i or $-i$ could be chosen to equal $\sqrt{-1}$, since i's only trait is that $i^2 = -1$. Therefore, any true statement about i is also true of $-i$. "It is difficult, once one has grasped this, to have any respect for the view that i might denote an independently existing Platonic object," Gowers writes.

By definition, though, the Platonic realm includes objects with split identities. As with counting numbers, the human mind contributed the terminology, but not the concept. Also by definition, anything that exists in the spatiotemporal realm exists in the non-spatiotemporal realm, including concepts such as justice and freedom and numbers with peculiar attributes.

. . .

ENDORSING PLATONISM MEANS allowing the possibility
of the world's being more capacious than we can conceive. Evolution might be progressing on a grand scale,
too. All that we are meant to know or might be capable
of knowing could be unfolding over a longer span than a
human life or even generations of lives can enclose. We
might not yet be equipped to think all that there is to be
thought.

When I read, under "Profession of Faith," in Simone
Weil's "Draft for a Statement of Human Obligation,"
that "there is a reality located outside the world, that is
to say, outside time and space, outside man's mental universe, outside the entire domain that human faculties can
reach," something inside me that, having no better word
for it, I will call my spirit responds.

6.

Mathematics, which settles all questions it is capable of
settling, and has plenty of pronouncements about real
and imaginary worlds, offers very little about where it
comes from. Mathematics is like the enthralling dinner
companion you speak to intimately all evening then realize as you rise from the table that all the revelations were
your own.

With mathematics it is impossible to avoid metaphysical questions. It begins, after all, with an unanswerable
question: Where do numbers come from? This is one
reason mathematics is bound up with spiritual explanations. To think about the origin of numbers is to allow
a metaphysical answer or at least not be able to exclude

it. It doesn't insist on it, but it accepts the possibility. I don't know if there is another subject that allows this that isn't already philosophy. We know when the universe began and how. We know how life came to be and how it refines itself through evolution. We don't know where numbers come from or why they have the properties they do, unless you believe that they are a system invented by humans based on the ways in which we apprehend the world, a creation of our thinking and therefore our neurology. In that case numbers and mathematics form a mechanical system that we expand according to rules we make up. While this is an assertion that I don't find persuasive, I cannot dismiss it, either.

I did not expect to be drawn into thinking about these kinds of things. I didn't even want to be. I thought I was merely setting out to do teenage math. It has made me have to consider in detail what I believe, which is taxing and never-ending, since, as with all of us, what I believe progresses, or at least changes, so long as I read and talk with others. To believe in anything is also to risk having the belief overthrown, which is not always pleasurable. It's a form of exposure.

The outcome of thinking, of self-consciousness, suggests that it is always possible to be on the way to becoming some other version of oneself. Does that mean that we are always discovering aspects of ourselves that we didn't know about or that we are instead creating new ones? Do we have potential selves? Are there an infinite number of them? Last self known + 1? Do we not arrive at the end of ourselves for the same reason we can't arrive at the end of numbers: we run out of time? Plato describes time as the "moving image of eternity." It is one of those occasions when an ancient mind seems startlingly modern.

7.

Around eleven weeks in, it occurred to me, later than it might have, that it would have helped me as an adolescent to know the theoretical basis of the work I was being asked to do, and that I was leaving arithmetic, which was mechanical, for algebra, which was conceptual. "I call x a number to be named later," Deane Yang told me. "Like you're trading baseball players, but at the moment, you don't know what number it is. So algebra is same rules as arithmetic, but you're dragging along a number to be determined later. You can't make it disappear like the $2 + 1$ you can make disappear and become 3. $x + 1$ won't disappear. That's the conceptual idea of algebra."

It was, I see now, the secret knowledge that the others had been let in on, although by whom and how I don't know. Or maybe it had been explained, and I hadn't paid close enough attention.

It was also borne in on me that something that works can't be proved not to work, and I would have to accept that the shortcoming was almost entirely my own. I hadn't been cheated. I hadn't been wronged. I *was* wrong. The fault was in my reactionary thinking. This was not fun to own up to, but it allowed me to loosen the grip of the embittered adolescent.

IDEAS DO NOT organize themselves for me easily; they change position and occasionally switch places. I sometimes think that I am intellectually dyslexic. I admire straightforward, logical thinking, but I can't do it, so I

don't like doing puzzles. Seeing puzzles as an opportunity
for defeat is one reason I grew frustrated at being balked
by algebra problems. Equations in adolescent algebra are
simple sentences, equivalent to the sentences in books for
young adult readers. To be unable to read a simple sentence correctly made me feel flat-out dense. What Deane
had been telling me, obliquely, was that I was reading
them too closely. They weren't each an occasion to be
reasoned out, they were invitations to evoke principles I
was expected to recognize and enact. They were generalizations. It was strange to become aware in early old age
that a temperamental circumstance, my tendency to look
too closely at things and be too subjective, also operated
in me as a boy and kept me from seeing what the teacher
and the textbooks had plainly intended that I see. I had felt
only the sense of knowledge withheld, of secretiveness, of
being excluded from the circle of initiates, and the adolescent imperative to conceal that I was lost.

FROM A JOURNAL, week twelve:

> I am fatigued by the number of rules, procedures,
> and formulas. I feel as if I am training for a profes-
> sion that I plainly won't be good at. Not grasping
> the degree to which generalizations are allowed
> only makes it harder to advance. I can think of
> no equivalent except the obvious one that Deane
> points out, grammar, but there is something nat-
> ural about learning spoken grammar—the ear
> listens and tends to adapt, without all that much
> study, as if we already knew some of what to do.

There is no mathematical equivalent I can think of. We don't go through childhood converting the flights of birds into vectors and parabolas, or parsing the passing of traffic, or the number of people we encounter and the intervals in which we encounter them compared to the hours that pass. Beyond counting and arithmetic, there is no natural way of learning mathematics. It's true that we perform a number of complex mathematical calculations effortlessly—we figure out how to catch a ball, we estimate whether waiting on a line is sensible—but they aren't occasions we address by consciously considering their mathematics.

It is staggering to make so many simple mistakes. In no other part of my life am I so inept, so far as I know. Simple competence eludes me.

I assumed in my second encounter that eventually I would get everything right, but that isn't the way it happened. I never achieved mastery in algebra or even, really, much proficiency. I was maybe a B–, C+ student. I had meant to enjoy prevailing. I imagined meeting algebra again as being like the encounter one has at a twentieth reunion with the high school quarterback who now does odd jobs and breaks out his wallet to show you the clippings of his best games.

I had hoped to solve problems by knowing the right methods and carrying them out infallibly. Sometimes I chose the right method and carried it out fallibly, so that the answer might be –4 and I had 4. Sometimes I began with the wrong method and ended with a term that was absurdly remote from the answer. Then I would go backward through the steps and was especially vexed if I got to where I had started without finding the wrong turn I had taken. What I had intended to enact was a ruthless and complacent accuracy, like what my bright classmates had shown. When the teacher placed the tests on our desks and said, "Begin," they picked up their pencils and lowered their heads and took off like trains intent on reaching their destinations on schedule or maybe even early. Copying their work, I found it difficult sometimes even to keep up with them.

After three months of algebra, I was feeling unsteady. I had expected to be pleased at having unraveled starter math's intricacies and using them the way I might be using new French to read *Le Monde* or Spanish to watch telenovelas. I had thought a world would be opening to admit me, and I guess it was, but only tepidly and in evening light, or as if I were stuck looking at the view from the back of the crowd. What was supposed to make me happy was making me unhappy. Cunning is what it seems to me that student math often requires, not thought. Thought is expansive. Cunning is opportunistic and narrow and strategic.

Despite my resolutions, I persisted in feeling that equations were rigged. When I examined them, I was looking not for how to solve them so much as where the ambush was. This was not an efficient state of mind. Receptivity was called for, not skepticism. When I was feeling that way, mathematics struck me as a species of fancy grifter.

I LIKE, THOUGH, that I am learning an ancient practice, that I am reading from "the book of dark or mysterious things." I sometimes feel a kindred quality to this in church, on the few times I am ever there, a sense that the procedures were established by antiquated figures and have been invoked the world over by uncountable numbers of people on uncountable occasions since. Mathematics, though, is older than church. One might reply, perhaps, that ideas of divinity are even older, but while they are entertained in the mind, they don't seem to be thoughts. They are responses, intuitions, and impressions, and somehow static and different from the dynamic quality of thinking, which revises itself constantly or it isn't thinking. Both the occult and the mathematical, though, have a degree of superstition in their upbringings, divinity through the contributions of spirit-thinkers and mathematics through Pythagoras.

I find myself annoyed at having to accept an illogical standard, the square root of 4, for example, being both 2 and −2. In *Notes from the Underground*, Dostoyevsky writes that mathematical certainty is an insufferable fact. "I admit that twice two makes four is an excellent thing," he writes, "but if we are to give everything its due, twice two makes five is sometimes a very charming thing too."

WEEK FOURTEEN: JUDGING from the biographical notes at the back of textbooks, a considerable number of math teachers prefer to be called educators. They also believe that learning algebra is a regular minefield of unexpected pleasures. I think it would have been pleasant for me not to have felt left behind. Being not educated was decidedly not fun. Finding even a beginner's way through the thicket of symbols and procedures might have given me a satisfaction and confidence I felt the lack of acutely. I see now that learning math wasn't entirely beyond my capabilities, but I didn't then, when it mattered more.

To defend myself as a boy against the assault that failing at math was causing, I had to feel that math was irrational and inconsistent. Or perhaps that a secret knowledge had been denied me. Or to think that I had no ability for the task. Now, equipped to untangle some of the enigmas, I see, as Amie has told me, that math is logical, and I feel indignant that this, and not something mysterious and enlarging, is what was withheld from me.

Amie's husband, Benson Farb, is also a mathematician, and once when I told her my feelings about math's contradictions, she said, "I can't tell you how many times Benson and I have said to each other over the years, 'I found a contradiction in math.'"

"There really are such things?"

"No, it's what you say when you don't get the answer you expect."

Mathematics might be the only creative pursuit in which inevitability figures. Other artists might be defeated by a task beyond their capabilities, but they do not live with knowing that sooner or later, if their work is consequential, someone will do what they haven't been able to do. Mathematicians work within a discipline in which, so long as their suppositions are correct, there is always a precise and irrefutable answer, even if they can't find it.

8.

I was interested to know what I could of how a pure mathematician thought and what it was like to inhabit an imaginary world. Amie suggested I speak to Yitang Zhang, a solitary part-time calculus teacher at the University of

New Hampshire who had received several prizes, including a MacArthur award, for solving a problem that had been open for more than a hundred and fifty years. His proof was deeply complex, she said, but its outlines were not so difficult.

The problem that Zhang chose, in 2010, is from number theory and is usually called "bounded gaps." It concerns prime numbers and whether there is a boundary within which, on an infinite number of occasions, two consecutive prime numbers can be found, especially out in the region where the numbers are so large that it would take a book to print a single one of them. Daniel Goldston, a professor at San Jose State University; János Pintz, a fellow at the Alfréd Rényi Institute of Mathematics, in Budapest; and Cem Yıldırım, of Boğaziçi University, in Istanbul, working together in 2005, had come close to establishing whether there might be a boundary, and what it might be. Goldston didn't think he'd see the answer in his lifetime. "I thought it was impossible," he told me.

Zhang, who also calls himself Tom, had published only one paper, to quiet acclaim, in 2001. In 2010, he was fifty-five. "No mathematician should ever allow himself to forget that mathematics, more than any other art or science, is a young man's game," Hardy wrote in *A Mathematician's Apology*. He also wrote, "I do not know of an instance of a major mathematical advance initiated by a man past fifty." Zhang had received a PhD in algebraic geometry from Purdue in 1991. His adviser, T. T. Moh, with whom he parted unhappily, wrote a description on his website of Zhang as a graduate student: "When I looked into his eyes, I found a disturbing soul, a burning bush, an explorer who wanted to reach the North Pole." Zhang left Purdue without Moh's support and, having published no papers, was unable to find an academic job. He lived, sometimes with friends, in Lexington, Kentucky, where he had occasional work, and in New York City, where he also had friends and occasional work. In Kentucky, he became involved with a group interested in Chinese democracy. Its slogan was "Freedom, Democracy, Rule of Law, and Pluralism." A member of the group, a chemist in a lab, opened a Subway franchise as a means of raising money. "Since Tom was a genius at numbers," a member of the group told me, "he was invited to help him." Zhang kept the books. "Sometimes, if it was busy at the store, I helped with the cash register," Zhang told me when I went to see him in New Hampshire. "Even I knew how to make the sandwiches, but I didn't do it so much." When Zhang wasn't working, he would go to the library at the University of Kentucky and read journals in algebraic geometry and number theory. "For years, I didn't really keep up my dream in mathematics," he said.

"You must have been unhappy."

He shrugged. "My life is not always easy," he said.

With a friend's help, Zhang eventually got his position in New Hampshire, in 1999. Having chosen bound gaps in 2010, he was uncertain of how to find a way into the problem. "I am thinking, Where is the door?" Zhang said. "In the history of this problem, many mathematicians believed that there should be a door, but they couldn't find it. I tried several doors. Then I start to worry a little that there is no door."

"Were you ever frustrated?"

"I was tired," he said. "But many times I just feel peaceful. I like to walk and think. This is my way. My wife would see me and say, 'What are you doing?' I said, 'I'm working, I'm thinking.' She didn't understand. She said,

'What do you mean?'" The problem was so complicated, he said, that "I had no way to tell her."

Deane Yang told me that a mathematician at the beginning of a difficult problem is "trying to maneuver his way into a maze. When you try to prove a theorem, you can almost be totally lost to knowing exactly where you want to go. Often, when you find your way, it happens in a moment, then you live to do it again."

ZHANG IS DEEPLY reticent, and his manner is formal and elaborately polite. Once when we were walking, he said, "May I use these?" He meant a pair of clip-on shades, which he held toward me as if I might want to examine them first. His enthusiasm for answering questions about himself and his work is slight. About half an hour after I had met him for the first time, he said, "I have a question." We had been talking about his childhood. He said, "How many more questions you going to have?" He depends heavily on three responses: "Maybe," "Not so much," and "Maybe not so much." From diffidence, he often says "we" instead of "I," as in, "We may not think this approach is so important." Occasionally, preparing to speak, he hums. After he published his result, he was invited to spend six months at the Institute for Advanced Study, in Princeton. The filmmaker George Csicsery has made a documentary about Zhang, called *Counting from Infinity*. In it, Peter Sarnak, a member of the Institute for Advanced Study, says that one day he ran into Zhang and said hello, and Zhang said hello, then Zhang said that it was the first word he'd spoken to anyone in ten days. Sarnak thought that was excessive even for a mathematician, and he invited Zhang to have lunch once a week.

Matthew Emerton, a colleague of Amie and Benson at the University of Chicago, also met Zhang at Princeton. "I wouldn't say he was a standard person," Emerton told me. "He wasn't gregarious. I got the impression of him being reasonably internal. He had received another prize, so the people around him were talking about that. Probably most mathematicians are very low-key about getting a prize, because you're not in it for the prize, but he seemed particularly low-key. It didn't seem to affect him at all." Deane attended three lectures that Zhang gave at Columbia in 2013. "You expect a guy like that to want to show off or explain how smart he is," he said. "He gave beautiful lectures, where he wasn't trying to show off at all." The first talk that Zhang gave on his result was at Harvard, before the result was published. A professor there, Shing-Tung Yau, had heard about Zhang's paper, and invited him. About fifty people showed up. One of them, a Harvard math professor, thought Zhang's talk was "pretty incomprehensible." He added, "The problem is that this stuff is hard to talk about, because everything hinges on some delicate technical understandings." Another Harvard professor, Barry Mazur, told me that he was "moved by his intensity and how brave and independent he seemed to be."

In New Hampshire, Zhang worked in an office on the third floor of the math and computer science building. His office had a desk, a computer, two chairs, a whiteboard, and some bookshelves. Through a window he looked into the branches of an oak tree. The books on his shelves had titles such as *An Introduction to Hilbert Space* and *Elliptic Curves, Modular Forms, and Fermat's Last Theorem*. There were also books on modern history and on Napoleon, who fascinates him, and copies of Shakespeare, which he reads in Chinese, because it's easier than Elizabethan English.

Eric Grinberg, a former chairman of the math department at the University of Massachusetts Boston, was a colleague of Zhang's in New Hampshire from 2003 to 2010. "Tom was very modest, very unassuming, never asked for anything," Grinberg told me. "We knew he was working on something important. He uses paper and a pencil, but the only copy was on his computer, and about once a month I would go in and ask, 'Do you mind if I make a backup?' Of course, it's all in his head anyway. He's above average in that."

Zhang's memory is abnormally retentive. A friend of his named Jacob Chi told me, "I take him to a party sometimes. He doesn't talk, he's absorbing everybody. I say, 'There's a human decency; you must talk to people, please.' He says, 'I enjoy your conversation.' Six months later, he can say who sat where and who started a conversation, and he can repeat what they said."

"I may think socializing is a way to waste time," Zhang says. "Also, maybe I'm a little shy."

A few years ago, Zhang sold his car, because he didn't really use it. He rented an apartment about four miles from campus and rode to and from his office with students on a school shuttle. He sits on the bus and thinks. Seven days a week, he arrives at his office around eight or nine and stays until six or seven. The longest he has taken off from thinking is two weeks. Sometimes he wakes in the morning thinking of a math problem he had been considering when he fell asleep. Outside his office is a long corridor that he likes to walk up and down. Otherwise, he walks outside.

Zhang met his wife at a Chinese restaurant on Long

Island, where she was a waitress. Her name is Yaling, but she calls herself Helen. A friend who knew them both took Zhang to the restaurant and pointed her out. "He asked, 'What do you think of this girl?'" Zhang said. Meanwhile, she was considering him. To court her, Zhang went to New York every weekend for several months. The following summer, she came to New Hampshire. She didn't like the winters, though, and moved to California, where she works at a beauty salon. She and Zhang have a house in San Jose, and he now lives in California and teaches at UC Santa Barbara.

Until Zhang was promoted to professor, as a consequence of his proof, his appointment had been tenuous. "I was chair of the math department, and I had to go to him from time to time and remind him this was not a permanent position," Eric Grinberg said. "We were grateful to him, but it's not guaranteed. He always said that he very much appreciated the time he had spent in New Hampshire."

ZHANG DEVOTED HIMSELF to bounded gaps for a couple of years without finding a door. "We couldn't see any hope," he said. Then, on July 3, 2012, in the middle of the afternoon, "within five or ten minutes, the way is open."

Zhang was in Pueblo, Colorado, visiting Jacob Chi, who is a music professor at Colorado State University Pueblo. A few months earlier, Chi had reminded Zhang that he had promised one day to teach his son, Julius, calculus, and since Julius was about to be a senior in high school Chi had called and asked, "Do you keep your promise?" Zhang spent a month at the Chis'. Each morning, he and Julius worked for about an hour. "He didn't have a set curriculum," Julius told me. "It all just flowed from his memory. He mentioned once that he didn't have any numbers in his phone book. He memorized them all."

Zhang had planned a break from work in Colorado, and hadn't brought any notes with him. On July 3, he was walking around the Chis' backyard. "We live in the mountains, and the deer come out, and he was smoking a cigarette and watching for the deer," Chi said. "No deer came," Zhang said. "Just walking and thinking, this is my way." For about half an hour, he walked around at a loss.

More or less suddenly, "I see numbers, equations, and something even—it's hard to say what it is," Zhang said. "Something very special. Maybe numbers, maybe equations—a mystery, maybe a vision. I knew that, even though there were many details to fill in, we should have a proof. Then I went back to the house."

Zhang didn't say anything to Chi about his breakthrough. That evening, Chi was conducting a dress rehearsal for a Fourth of July concert in Pueblo, and Zhang went with him. "After the concert, he couldn't stop humming 'The Stars and Stripes Forever,'" Chi said. "All he would say was, 'What a great song.'"

9.

I asked Zhang, "Are you very smart?" and he said, "Maybe, a little." He was born in Shanghai in 1955. His mother was a secretary in a government office, and his father was a college professor whose field was electrical engineering. As a small boy, he began "trying to know everything in mathematics," he said. "I became very thirsty for math." His parents moved to Beijing for work, and Zhang remained in Shanghai with his grandmother. The revolution had closed the schools. He spent most of his time reading math books that he ordered from a bookstore for less than a dollar. He was fond of a series whose title he translates as "A Hundred Thousand Questions Why." There were volumes for physics, chemistry, biology, and math. When he didn't understand something, he said, "I tried to solve the problem myself, because no one could help me."

Zhang moved to Beijing when he was thirteen, and when he was fifteen he was sent with his mother to the countryside, to a farm, where they grew vegetables. His father was sent to a farm in another part of the country. If Zhang was seen reading books on the farm, he was told to stop. "People did not think that math was important to the class struggle," he said. After a few years, he returned to Beijing, where he got a job in a factory making locks. He began studying to take the entrance exam for Peking University, China's most respected school: "I spent several months to learn all the high school physics and chemistry, and several to learn history. It was a little hurried." He was admitted when he was twenty-three. "The first year, we studied calculus and linear algebra—it was very exciting," Zhang said. "In the last year, I selected number theory as my specialty." Zhang's professor insisted, though, that he change his major to algebraic geometry, his own field. "I studied it, but I didn't really like it," Zhang said. "That time in China, still the idea was like this: the individual has to follow the interest of the whole group, the country. He thought algebraic geometry was more important than number theory. He forced me. He was the university president, so he had the authority."

During the summer of 1984, T. T. Moh visited Peking University from Purdue and invited Zhang and several other students, recommended to him by Chinese professors, to do graduate work in his department. One of Moh's specialties is the Jacobian conjecture, and Zhang was eager to work on it. The Jacobian conjecture, a problem in algebraic geometry that was introduced in 1939 and is still unsolved, stipulates certain simple conditions that, if satisfied, enable someone to solve a series of complicated equations. It is acknowledged as being beyond the capacities of a graduate student and approachable by only the most accomplished algebraic geometers. A mathematician described it to me as a "disaster problem," for the trouble it has caused. For his thesis, Zhang submitted a weak form of the conjecture, meaning that he attempted to prove something implied by the conjecture, rather than to prove the conjecture itself.

After Zhang received his doctorate, he told Moh that he was returning to number theory. "I was not the happiest," Moh wrote me. "However, I was for the student's right to change fields, so I kept my smile and said bye to him. For the past 22 years, I knew nothing about him."

After graduating, most of the Chinese students went into either computer science or finance. One of them, Perry Tang, who had known Zhang in China, took a job at Intel. In 1999, he called Zhang. "I thought it was unfair for him not to have a professional job," Tang told me. He and Zhang had a classmate at Peking University who had become a professor of math at the University of New Hampshire, and when the friend said that he was looking for someone to teach calculus Tang recommended Zhang. "He decided to try him at a temporary position," Tang said.

10.

Zhang finished "Bounded Gaps Between Primes" in late 2012; then he spent a few months methodically checking each step, which he said was "very boring." On April 17, 2013, without telling anyone, he sent the paper to *Annals of Mathematics*, widely regarded as the profession's most prestigious journal. In the *Annals* archives are unpublished papers claiming to have solved practically every math problem that anyone has ever thought of, and others that don't really exist. Some are from people who "know a lot of math, then they go insane," a mathematician told me. Such people often claim that everyone else who has attacked the problem is wrong. Or they announce that they have solved several problems at once, or "they say they have solved a famous problem along with some unified field theory in physics," the mathematician said. Journals such as *Annals* are skeptical of work from someone they have never heard of.

In 2013, *Annals* received nine hundred and fifteen papers and accepted thirty-seven. The wait between acceptance and publication is typically around a year. When a paper arrives, "it is read quickly, for worthiness," Nicholas Katz, the Princeton professor who is the journal's editor, told me, and then there is a deep reading that can take months. "The paper I can't evaluate off the top of my head, my role is to know whom to ask," Katz said. "In this case, the person wrote back pretty quickly to say, 'If this is correct, it's really fantastic. But you should be careful. This guy posted a paper once, and it was wrong. He never published it, but he didn't take it down, either.'" The reader meant a paper that Zhang posted on the website arXiv.org, where mathematicians often post results before submitting them to a journal, in order to have them seen quickly. Zhang posted a paper in 2007 that fell short of a proof. It concerned another famous problem, the Landau-Siegel zeros conjecture, and he left it up because he hopes to correct it.

Katz sent "Bounded Gaps Between Primes" to a pair of readers, who are called referees. One of them was Henryk Iwaniec, a professor at Rutgers, whose work was among that which Zhang had drawn on. "I glanced for a few minutes," Iwaniec told me. "My first impression was: so many claims have become wrong. And I thought, I have other work to do. Maybe I'll postpone it. Remember that he was an unknown guy. Then I got a phone call from a friend, and it happened he was also reading the paper. We were going to be together for a week at the Institute for Advanced Study, and the intention was to do other work, but we were interrupted with this paper to read."

Iwaniec and his friend, John Friedlander, a professor at the University of Toronto, read with increasing attention. "In these cases, you don't read A to Z," Iwaniec said. "You look first at where is the idea. There had been nothing written on the subject since 2005. The problem was too difficult to solve. As we read more and more, the chance that the work was correct was becoming really great. Maybe two days later, we started looking for completeness, for connections. A few days passed, we're checking line by line. The job is no longer to say the work is fine. We are looking to see if the paper is truly correct."

After a few weeks, Iwaniec and Friedlander wrote to Katz, "We have completed our study of the paper 'Bounded Gaps Between Primes' by Yitang Zhang." They went on, "The main results are of the first rank. The author has succeeded to prove a landmark theorem in the

distribution of prime numbers." And, "Although we studied the arguments very thoroughly, we found it very difficult to spot even the smallest slip. . . . We are very happy
to strongly recommend acceptance of the paper for publication in the *Annals*."

When Zhang heard from *Annals*, he called his wife
in San Jose. "I say, 'Pay attention to the media and news-
papers,'" he said. "'You may see my name,' and she said,
'Are you drunk?'"

11.

"Bounded Gaps Between Primes" is a backdoor attack
on the twin-prime conjecture, which was proposed in
the nineteenth century and says that, no matter how far
you travel on the number line, even as the gap widens between primes, you will always encounter a pair of primes
that are separated by two, such as 5 and 7. The twin-
prime conjecture is still unsolved. Zhang established that
there is a distance within which, on an infinite number of
occasions, there will always be two primes.

"You have to imagine this coming from nothing," Eric
Grinberg said. "We simply didn't know. It is like thinking that the universe is infinite, unbounded, and finding
it has an end somewhere." Picture it as a ruler that might
be applied to the number line. Zhang chose a ruler of a
length of seventy million, because a number that large
made it easier to prove his conjecture. (If he had been
able to prove the twin-prime conjecture, the number for
the ruler would have been two.) This ruler can be moved
along the line of numbers and enclose two primes an

infinite number of times. Something that holds for infinitely many numbers does not necessarily hold for all. For example, an infinite number of numbers are even, but an infinite number of numbers are not even, because they are odd. Similarly, this ruler can also be moved along the line of numbers an infinite number of times and not enclose two primes.

From Zhang's result, a deduction can be made, which is that there is a number smaller than seventy million which precisely defines a gap separating an infinite number of pairs of primes. You deduce this, Amie told me, by means of the pigeonhole principle. You have an infinite number of pigeons, which are pairs of primes, and you have seventy million holes. There is a hole for primes separated by two, by three, and so on. Each pigeon goes in a hole. Eventually, one hole will have an infinite number of pigeons. It isn't possible to know which one. There may even be many, there may be seventy million, but at least one hole will have an infinite number of pigeons.

Having discovered that there is a gap, Zhang wasn't interested in finding the smallest number defining the gap. This was work that he regarded as a mere technical problem, a type of manual labor—"ambulance chasing" is what a prominent mathematician called it. Nevertheless, within a week of Zhang's announcement mathematicians around the world began competing to find the lowest number. One of the observers of their activity was Terence Tao, who had the idea for a cooperative project in which mathematicians would work to lower the number rather than "fighting to snatch the lead," he told me.

The project, called Polymath8, started in March of 2013 and continued for about a year. Incrementally, relying also on work by a young British mathematician named

James Maynard, the participants reduced the bound to 246. "There are several problems with going lower," Tao said. "More and more computer power is required— someone had a high-powered computer running for two weeks to get that calculation. There were also theoretical problems. With current methods, we can never get better than six, because of something called the parity problem, which no one knows how to get past." The parity problem says that primes with certain behaviors can't be detected with current methods. "We never strongly believed we would get to two and prove the twin-prime conjecture, but it was a fun journey," Tao said.

12.

"Is there a talent a mathematician should have?"

"Concentration," Zhang said. We were walking across the campus in a light rain. "Also, you should never give up in your personality," he continued. "Maybe something in front of you is very complicated, it's lengthy, but you should be able to pick up the major points by your intuition."

When we reached Zhang's office, I asked how he had found the door into the problem. On a whiteboard, he wrote, "Goldston-Pintz-Yıldırım" and "Bombieri-Friedlander-Iwaniec." He said, "The first paper is on bound gaps, and the second is on the distribution of primes in arithmetic progressions. I compare these two together, plus my own innovations, based on the years of reading in the library."

When I asked Peter Sarnak how Zhang had arrived at

his result, he said, "What he did was look way out of reach. Maybe forty years ago the problem appeared hopeless, but in 2005 Goldston-Pintz-Yıldırım put it at the doorstep. Everybody thought, Now we're very close, but by 2011 no one was making any progress. Bombieri, Friedlander, and Iwaniec had the other important work, but it looked like you couldn't combine their ideas with Goldston. Their work was just not flexible enough to jive—it insisted on some side conditions. Then Zhang comes along. A lot of people use theorems like a computer. They think, If it is correct, then good, I'll use it. You couldn't use the Bombieri-Friedlander-Iwaniec, though, because it wasn't flexible enough. You have to take my word, because even to a serious mathematician this would be difficult to explain. Zhang understood the techniques deeply enough so as to be able to modify Bombieri-Friedlander-Iwaniec and cross this bridge. This is the most significant thing about what he has done mathematically. He's made the Bombieri-Friedlander-Iwaniec technique about the distribution of prime numbers a tool for any kind of study of primes. A development that began in the eighteen-hundreds continued through him."

"Our conditions needed to be relaxed," Iwaniec told me. "We tried, but we couldn't remove them. We didn't try long, because after failing you just start thinking there is some kind of natural barrier, so we gave up."

I asked if he was surprised by Zhang's result. "What Zhang did was sensational," he said. "His work is a masterpiece. When you talk of number theory, a lot of the beauty is the machinery. Zhang somehow completely understood the situation, even though he was working alone. That's how he surprised. He just amazingly pushed further some of the arguments in these papers."

Zhang used a very complicated form of a simple device for finding primes called a sieve, invented by Eratos-
thenes, a contemporary of Archimedes. To use a simple
sieve to find the primes less than a thousand, say, you
write down all the numbers, then cross out the multiples
of two, which can't be prime, since they are even. Then
you cross out the multiples of three, then five, and so on.
You have to go only as far as the multiples of thirty-one.
Zhang used a different sieve from the one that others had
used. The previous sieve excluded numbers once they
grew too far apart. With it, Goldston, Pintz, and Yıldırım
had proved that there were always two primes separated
by something less than the average distance between
primes that large. What they couldn't identify was a precise gap. Zhang succeeded partly by making the sieve less
selective.

13.

I asked Zhang if he was working on something new.
"Maybe two or three problems I would like to solve," he
said. "Bounded gaps is successful, but still I have something else."

"Will it be as important?"

"Yes."

According to other mathematicians, Zhang is working
on his incomplete result for the Landau-Siegel zeros conjecture. "If he succeeds, it would be much more dramatic,"
Peter Sarnak said. "We don't know how close he is, but
he's proved that he's a genius. There's no question about
that. He's also proved that he can speak with something

over many years. Based on that, his chances are not zero. They're positive."

"Many people have tried that problem," Iwaniec said. "He's a private guy. Nothing is rushed. If it takes him another ten years, that's fine with him. Unless you tackle a problem that's already solved, which is boring, or one whose solution is clear from the beginning, mostly you are stuck. But Zhang is willing to be stuck much longer."

Zhang's preference for undertaking only ambitious problems is rare. The pursuit of tenure requires an academic to publish frequently, which often means refining one's work within a field, a task that Zhang has no inclination for. He does not appear to be competitive with other mathematicians, or resentful about having been simply a teacher for years while everyone else was a professor. No one who knows him thinks that he is suited to a tenure-track position. "I think what he did was brilliant," Deane told me. "If you become a good calculus teacher, a school can become very dependent on you. You're cheap and reliable, and there's no reason to fire you. After you've done that a couple of years, you can do it on autopilot; you have a lot of free time to think, so long as you're willing to live modestly. There are people who try to work nontenure jobs, of course, but usually they're nuts and have very dysfunctional personalities and lives, and are unpleasant to deal with, because they feel disrespected. Clearly, Zhang never felt that."

One day, I arrived at Zhang's office as he was making tea. There was a piece of paper on his desk with equations on it and a pen on top of the paper. Zhang had an envelope in one hand. "I had a letter from an old friend," he said. "We have been separated for many years, and now he found me."

He took a pair of scissors from a drawer and cut open
the envelope so slowly that he seemed to be performing
a ritual. The letter was written in Chinese characters. He
sat on the edge of his chair and read slowly. He put the
letter down and took from the envelope a photograph of
a man and a woman and a child on a sofa with a curtain
in the background. He returned to reading the letter,
and then he put it back in the envelope and in the drawer
and closed the drawer. "His new address is in Queens," he
said. Then he picked up his tea and blew on it and faced
me, looking at me over the top of the cup like someone
peering over a wall.

I asked about Hardy's observations regarding age—
Hardy also wrote, "A mathematician may still be competent enough at sixty, but it is useless to expect him to have
original ideas."

"This may not apply to me," Zhang said. He put his
tea on the desk and looked out the window. "Still I am
confident of myself," he said. "Still I have some other
visions."

14.

My adult life has been framed by the attempt to find
words to convey my experience to another person. One
of the difficulties is writing a sentence that means what
I intend it to mean. If you think of the number of ways
in which an actor might recite a line, then you recognize
the ways in which a sentence can be interpreted. Numbers are precise in a way that words can't be, but numbers
cannot explain my feelings for my wife or my son or how

rain feels or what it is like to stand beside the ocean in a storm.

It isn't that numbers are incapable of describing the world of feelings, it is more as if they have the capacity to represent a parallel experience, the way psychedelic drugs or alcohol makes a person aware of a parallel reality. ("The sway of alcohol over mankind is unquestionably due to its power to stimulate the mystical faculties of human nature," William James writes in *The Varieties of Religious Experience*, published in 1902.) With numbers, though, the reality exists and represents an analogous means of taking the measure of the world. Rain is no longer a sensation or an accessory to one's mood but a measurement of how much rain fell or how quickly or with what force. Quantitative implies time, another means of measuring sensation: when this amount of rain falls, I feel this way. The precision conveys meaning. I can find patterns in my surroundings the same way that I can find patterns among my impressions. The impressions run alongside the precise ones and sometimes they intersect, as in the case, say, of considering why the universe is structured in this or that way.

The finding of a formula for predicting the arrival of primes, if there is one, might not teach us anything that we don't already know. But what if it did? What if other things became apparent in the world's design because we understood this antique enigma? What else are primes linked to? Is it workmanlike knowledge or pure art? Or something else, I don't know what. Would it imply a divinity? And in what sense a divinity?

15.

Week fifteen: When the steps aren't obvious, I find it difficult to keep in mind all of an equation's terms. Knowing when to apply which steps has become a torment, since some equations that seem to involve dedicated procedures also don't seem to always. I feel a kinship with an autistic boy I read about who said, "I understand that people cry when they're sad, but I'm told that people also cry when they're happy. How am I supposed to know the difference?"

More than once and maybe more than twice, I have felt like an immigrant in a country whose language I don't entirely comprehend, unassimilated, sitting at the dinner table with my children who are fluent, saying things that are funny to them and not funny to me. I feel only the restrictions on my ability to speak. Marooned among math problems, I feel that I have a language, but a defective one.

To begin a day learning math, I have to prepare my mind, so as not to let my resistances overtake me. I know that I have to find the, apparently for me, elusive consistencies, the places where things behave as I expect them to, and where I am just employing methods to complete a task, where my textbook is a manual and not an inscrutable document.

16.

Now this: beauty has been largely dismissed in the humanities, since historical standards no longer persuade, but it still figures in mathematics. Mathematicians use the word beauty without irony or hesitation and differently from the way it is used in aesthetics. For mathematicians beauty is more a quality than an appraisal. Bertrand Russell writes, "Mathematics, rightly viewed, possesses not only truth, but supreme beauty—a beauty cold and austere, like that of sculpture, without appeal to any part of our weaker nature, without the gorgeous trappings of painting or music, yet sublimely pure, and capable of a stern perfection."

The German mathematician Hermann Weyl writes, "Mathematics has the inhuman quality of starlight, brilliant and sharp, but cold." Shing-Tung Yau, the professor at Harvard who invited Zhang to speak, told me that beauty was one of the most important elements of mathematics. "Otherwise it would be very boring and then you would not be able to stand it," he said.

Beauty does not depend on a correspondence with taste; it tends to make its claim in the interval before the mind focuses itself, and it is more easily felt than decided on. For mathematicians notions of mathematical beauty accord with a belief that the laws of the natural world, which mathematics frequently articulates, and the relations of whatever objects or circumstances mathematics is describing have a harmony, whether in the way that water moves or the orbits of the planets or the operations of shapes too complicated to see. Beauty in mathematical explanations is often a reflection of something glimpsed of a structure that seems to fit the world, although there is no explanation for why it should. Something, furthermore, that has been present but undetected and perhaps even unsuspected, the theory of relativity, for example. It illuminates a territory where the seen and unseen meet. This is a Pythagorean notion, never given up.

The philosopher Michael B. Foster, who taught at Oxford and died in 1959, believed that Christianity brought modern science into being, since it made the natural world worth examining for evidence of God's plan. During the Renaissance, mathematics began to be regarded as able to provide confirmation of God's gestures and thoughts by making hidden structures apparent. By the seventeenth century there is a near ecstatic blend of mathematics and the divine. This is exemplified by Descartes, who believed that God was a mathematician, and later, Newton, who said that space was the sensorium of God, meaning, I think—the remark has had a number of interpretations and is thought by some philosophers to be nonsense—that it was the territory in which God existed and operated and was itself a divine embodiment. This seems similar to me to Kepler's notion of the sphere as a godly metaphor. Newton also said, "Every newly found truth, every experiment or theorem, is a new mirror of the beauty of God," which is a quasi-Platonic remark.

Neuroscientists in Great Britain discovered that the same part of the brain that responds to art and music was activated in the brains of mathematicians when they looked at mathematics they thought beautiful. This finding was published, in the journal *Frontiers in Human Neuroscience*, in the article "The Experience of Mathematical Beauty and Its Neural Correlates." The mathematicians were asked to rate equations as ugly, neutral, or beautiful. Best of all they liked Leonhard Euler's identity $e^{i\pi} + 1 = 0$, "which links five fundamental mathematical constants with three basic arithmetic operations, each occurring once." Sir Michael Atiyah, a mathematician and one of the study's authors, told an interviewer, "It involves π; the mathematical constant e; . . . i, the imaginary unit; 1; and 0—it combines all the most important things in mathematics in one formula." Atiyah regarded it as the mathematical equivalent of "To be, or not to be." While the equation made use of only five symbols, it encapsulated "beautifully deep ideas," Atiyah said, "and brevity is an important part of beauty."

The equation that the mathematicians rated often as ugly and liked least was Srinivasa Ramanujan's infinite series for $1/\pi$,

$$\frac{1}{\pi} = \frac{2\sqrt{2}}{9801} \sum_{k=0}^{\infty} \frac{(4k)!\,(1103 + 26390k)}{(k!)^4 396^{4k}}$$

which I find strangely beautiful, in an elaborate, architectural sort of way. I like it also because the right-hand side of the equation seems like the inflated image that the left-hand side privately has of itself.

According to Hardy, a mathematical assertion is serious when it embodies significant ideas. What appears to be essential are "a certain generality and a certain depth." Generality means that an idea should be "constituent in many mathematical constructs," meaning useful in many ways and contexts. It also ought to be capable of being expanded to fit larger situations, and it should exemplify a class of theorems. Moreover, it should connect ideas that hadn't been connected. Theorems that are too specific, that don't illuminate larger ideas, various other mathematical peculiarities such as 8,712 and 9,801 being the only numbers smaller than 10,000 that are also "the only four figure numbers which are integral multiples of their 'reversals': 8712 = 4 × 2178 and 9801 = 9 × 1089" are only "odd facts, very suitable for puzzle columns and likely to amuse amateurs, but there is nothing in them which appeals much to a mathematician." Their proofs are "a little tiresome," and can't be generalized or made into what Hardy calls a "high-class theorem." A too-general theorem, however, is insipid. Mathematical ideas for Hardy are alive the way people are and "become dim unless they have plenty of individuality."

Generality is partly a matter of the ability to provide order. Mathematicians find pleasure in establishing order where there had not been order before, what Jean-Pierre Changeux describes as extracting "structure and invariance from the midst of disarray and turmoil." Often they achieve this by finding relations between or among unrelated fields. Poincaré said in 1908 that "the mathematical facts worthy of being studied" are "those which reveal to us unsuspected kinship between other facts, long known, but wrongly believed to be strangers to one another." Deane Yang told me that "mathematicians want for there to be two different universes, then you find a door between them, and all of a sudden they're one universe. Then you want to explore all the ways to get from one to the other."

Bertrand Russell liked thinking about deep mathematics because what was not human broadened one's sense of what *was* human. "The discovery that our minds are capable of dealing with material not created by them, above all, the realisation that beauty belongs to the outer world as to the inner," he writes, "are the chief means of overcoming the terrible sense of impotence, of weakness, of exile amid hostile powers."

The older I get the more I find this observation sustaining.

17.

The classroom, also week fifteen: Deane said I needed to understand logic, because math is unforgiving of logical errors. Math often left things out, to avoid being "too verbose." To avoid going forward blindly, I had to know what was missing. (This is different from math textbook writers leaving things out. Mathematicians leaving out unnecessary things are being concise. Textbook writers leaving out necessary things are being coy.)

"To have the exact intended meaning, every mathematical statement and equation has to be a complete, grammatically correct sentence," he said. "Focusing on mathematical grammar will force you to understand the meaning of what you are being asked to solve. You'll have to work more slowly for a while, but you'll be able to go faster once you know what you're doing."

By using correct grammar I took him to mean that I had to follow carefully the progress of each stage of an equation and not just quickly perform half of the equation, usually the easier half—a phrase, that is—then hastily transfer my solution into a following step without allowing completely for how the solution affected the meaning of the equation. If I felt lost at confronting a variable, I could insert a number in its place. "Professional mathematicians do this all the time, but it's rarely taught to students," he said. As it happened, I had already stumbled across this tactic, as a DIY means of demystifying equations. A lot of things I did in algebra I did in the spirit of exposing its irrationality, that is, destructively. I had a bad attitude. I was a type of math punk. Out of frustration, I sometimes inserted numbers in the place of variables, thinking, Show me how this is wrong, too, I dare you, and understood nothing when it wasn't.

It is important to learn algebra, Deane said, but more useful is learning the procedures that support it. I might begin by learning how to calculate the distance between two points by constructing a right triangle and using the Pythagorean theorem, and from there move to the equations defining the radius and diameter of a circle and satisfy myself as to why they work. When I reach a place where I don't understand what I'm doing, I should put the work aside to learn what else I needed. If I came to his classroom, he would give me problems and insist that I solve them sequentially, with no leaving one step for another until I had persuaded him that I knew why I could leave. I was aware that this was a kind and generous offer and sensible, too, but I didn't want to go to math camp. Furthermore, I didn't want to struggle in front of a witness and demonstrate that I might not be very bright, so for a while I avoided him.

Deane and Amie each have their biases. Amie wonders why I would bother learning word problems. Deane wonders why, while studying algebra, I would bother learning the quadratic formula, which is $x = \dfrac{-b \pm \sqrt{b^2 - 4ac}}{2a}$ and useful for finding solutions to quadratic equations. Partly I learned it because it was a shiny object that drew my attention. Also, because while it looks complicated, it isn't difficult to use. It is simply a matter of knowing values for a, b, and c. Squaring b and subtracting 4 times the product of ac and taking its square root then subtracting b and dividing it by 2 times a is something I could have done even before I started math again. Using it made me feel in a playacting way that I was doing complicated operations with numbers. Moreover, it was one of the few things I recognized already from its name, so it flattered me to have possession of it. Mainly, though, I thought it figured importantly in calculus, but he said it doesn't.

So far as I can tell, there are two ways that math is taught. The more common method insists on the inviolability of the material—it has its own rationale, which you must apprehend and submit to. The less common approach is that of trying to understand why a student finds learning math hard. As a boy I saw in mathematics only deceptions and tricks, so I searched for inconsistencies and grew resentful, bound to happen. If math was consistent, how come I couldn't learn it? I was anyway subject always to the first method, which, in the service of efficiency, is willing to shed you if you falter. Only so much time can be spent before a class grows restless, and the caravan folds its tents and moves on.

18.

I finished algebra plagued by the feeling that I had to get every problem right. I had felt lost as a boy in geometry and in Algebra II, but that was because I never found a foothold, and I had never expected to. With algebra, I had started hopefully and been throttled. What I had wished for on my second engagement was to see algebra as rational and cohesive, and therefore benign, so that I could dispose of the mystery it had left me with. If I were able to do that, I would have made use of ways of thinking that challenged me to expand my, shall I call it, my intellect? My capacity for regarding problems whose solutions require the management of symbols, something I had never been good at. I had always been fond of Hemingway's saying that in *The Old Man and the Sea*, the man was a man, and the fish was a fish. I might have gone the rest of my way through life believing that objects are only what they appear to be, if Amie hadn't told me of Maxwell's remark, which left me wondering what there was of the world to know that I didn't know, and what might be close enough at hand to be apprehended and made use of, even if awkwardly at first.

The enlargement of one's intellectual reach isn't the kind of circumstance a person can identify empirically. One can only sense it about oneself. I felt I was beginning to change, to a degree, perhaps only in a cursory way, but I also felt, superstitiously, that to acknowledge it might be prideful, which might lead to its being revoked by whatever agency it is that lurks inside superstitious moral attitudes. The Sunday school teacher who tells you that God sees everything and unsettles you about whether you are safe from notice as you lie in bed in the dark trying not to have thoughts you were told you shouldn't have. Anyway, I finished algebra, I came to the end of the textbook. It had taken five months, not six weeks. I had learned things, though, I had some new skills, even if rudimentary ones. I could do things I hadn't been able to do, and I was pleased. The accomplishment was not substantial, but it was my own, and I had worked for it. I raised a private glass to myself and said, "Well done," in the middle of an afternoon.

19.

All the mathematicians I knew told me that I would like geometry. Algebra was necessary and tolerable, but geometry had pleasures, they said. They practically insisted on it. I wonder if this attitude is partly sentimental. Algebra's origins are opaque. No one can place it precisely in history, and al-Jabr is a remote and not easily imagined figure, and on top of that an Arab, so all the prejudices of Western culture in favor of itself come down on his head, whereas geometry has the grand old man, Euclid, the Santa Claus of axioms. Geometry lends itself to a narrative. Algebra's past seems as diffuse as algebra is abstract.

Geometry also has a definitive text representing a contribution to culture that few men or women have ever made so concisely, perhaps no one else has. People writing about Euclid's *Elements* often like to point out that it is the second-best-selling book in the Western world, after the Bible. I would like to point out that it is strangely private and interior. Only a severely gifted mind, I think, could have addressed simple structures with such fanatic specificity.

Plato said, "Geometry is knowledge of what always is. It draws the soul towards truth, and produces philosophical thought." Euclid lived a generation after Plato and heard more arguments and geometrical truths than Plato had, and he wrote them down. In *Elements*, he is a compiler of historical and contemporary work. His book is a compendium of immortal remarks concerning the natural world, but it is generally agreed that only some of the remarks, and maybe almost none of them, are his own. His original work on mathematics he published in *Optics* and *Phaenomena*.

In *Elements* each entry leads to the next. It is regarded as a manual for how to think logically, and, so far as anyone knows, it is the first one that Western culture produced. Its methods of reasoning are adaptable to any field that is subject to procedural thinking. Abraham Lincoln, while a lawyer in Illinois, spent a year reading Euclid in order to teach himself to reason cogently. From Euclid someone can learn to construct an unassailable assertion. According to Stanislas Dehaene, Euclid makes explicit "the difference between if and if and only if."

Euclid was describing the attributes of forms when the world was new or at least hadn't had so many eyeballs on it. His was the period when the Greeks were naming the sky, and it must have seemed as if all creation were laid out for them to catalog and describe, and that there was an essential and beautiful order to the world. "Euclid alone has looked on Beauty bare," Edna St. Vincent Millay writes.

A STRIKING THING about geometry is that it begins with the acceptance of a point, and only one straight line that passes through two points, but whereas a line is real, a point is imaginary. A point is a location, and there are more of them in the universe than a number can be written to enumerate, but no one can find a point, because it has no width, height, or volume, and therefore no area. ("A point is that which has no part," Euclid says.) It's an idea accepted as a thing and is the simplest example I can think of to make one aware of how mathematics is both real and not real. Mathematics is rigorous largely because of Euclid and his insistence of proof, yet Euclidean geometry begins with a conceptual object being taken for an

actual one. This seems to me a fantastically strange circumstance, not to say another example of math's inconsistencies, but everyone else seems okay with it.

Euclid begins by assuming five propositions that
appear indisputable. He calls them postulates. They are
perhaps sufficiently well known that I needn't list them,
but even so, here they are, according to Thomas Heath's
translation, published in 1909 and regarded as definitive:

Let the following be postulated:

1. To draw a straight line from any point to any
   point.
2. To produce a finite straight line continuously in a
   straight line.
3. To describe a circle with any centre and distance.
4. That all right angles are equal to one another.
5. That, if a straight line falling on two straight lines
   make the interior angles on the same side less
   than two right angles, the two straight lines, if
   produced indefinitely, meet on that side on which
   are the angles less than the two right angles.

Postulate one means that a straight line can be drawn
between any two points. Technically, this is a line segment; a line continues infinitely in two directions. Postulate two means that a line segment can become a line
by being extended. Postulate three means that from a
straight line segment a circle can be drawn with the segment as its radius; one end point is the center of the circle.
Postulate four is probably plain enough, all right angles
are ninety degrees. Postulate five says that two lines, if a
third line is drawn across them, will intersect eventually

on the side on which the angles created by the third line are each less than ninety degrees. This is called the parallel postulate, since if the angles are ninety degrees, the lines are parallel.

The first four assertions unfold from one another, but the fifth does not and is unprovable. The belief that it might be proved was given up in the nineteenth century when non-Euclidean geometries were discovered. With these figures, the parallel postulate does not hold. On a sphere, for example, lines can be drawn that have a common perpendicular, as parallel lines do in the Euclidean sense, but they also intersect. Also, on a sphere, the shortest distance between two points is not always a straight line, which is why long airplane trips follow arcs.

Mathematicians have treated geometric forms, especially the circle, as species of natural history, as objects, that is, with attributes and inflexible behaviors. By Olympiodorus, a philosopher of the sixth century, the circle was regarded as a metaphor for the soul, "because it seeks itself, and is itself sought, finds itself and is itself found." When Ms. Scharfenstein, my teacher in second grade, said that it is impossible to draw a perfect circle, I don't think she knew that she was invoking Plato. I was at the age where such a remark seemed a challenge, and for a while I tried. With a protractor I found in a drawer in my father's workshop I drew circles that looked perfect and gave me an obscure sense of completion, of enclosing a space but also of the enclosure's being somehow endless, since a completed circle has no end point and therefore an everlasting continuity. It is the simplest object I can think of that is also beautiful. It is easy to see why it fascinated ancient thinkers so thoroughly, being both enclosure and boundary, finite within itself and infinite beyond its border. Celtic fishermen used sometimes to recite an invocation that included the lines "God be my unfolding / God be my circle."

20.

About infinity: Geometry implicitly addresses infinity, since a line has no end. The simplest description of infinity is a > b, although it is only an allusion. So far as I can discover, the concept of infinity first appears among the Jainists, a schismatic Hindu sect, in the sixth century BCE. The Jains believed that calculating with large numbers expanded a person's awareness. They had methods for estimating the time that a soul would need to complete its journey, and they had five descriptions of infinity: infinite in one and two directions, infinite size, infinite everywhere, and infinite perpetually, which read like a prayer.

The Greeks acknowledged two types of infinities, complete and incomplete, which are also called actual and potential. A complete infinity is a collection such as the set of whole numbers. An incomplete infinity is a series such as one that begins 1, 2, 3, . . . The first is specified by a concept in which all the terms are identified. The second accepts infinity as a term that is something like a direction toward which a collection heads. The first is bounded; the second is not. No matter where you are in either sequence, the end is never in sight.

Aristotle accepted only incomplete infinities. Among the Scholastics, whose ideas were taken from Aristotle's and were widely taught in Europe between the twelfth and eighteenth centuries, a completed infinity involved the "annihilation of number." Whereas with any whole number $a + b = c$, a + infinity = infinity, meaning a disappeared. The Greeks saw infinity as a boundary that could never be reached. The Scholastics saw a complete infinity as challenging God's omnipotence, since only God was infinite. Augustine believed not only that God was infinite, but that He was also absolute, meaning that He contained infinity. Thinking about infinity led one toward Him. In *Leaders of the Reformation*, published in London in 1860, John Tulloch quotes Martin Luther, sounding piqued in a dispute at a conference in 1529, saying, "I will have nothing to do with your mathematics! God is above mathematics!"

The early Greeks, those of, say, the fifth century BCE, imagined infinity as boundless. Aristotle, in the fourth century BCE, believed that something's being without bound was a specious idea, since, by definition, there could not be a body without boundaries. Infinity could exist only potentially, and it couldn't be demonstrated.

A companion theological belief is that the capacity to appreciate infinity was given to human beings by God. Being finite, we wouldn't have been able to conceive of infinity on our own. God gave us the means so that we could better appreciate His works. For Augustine, this proved God's existence.

Pure mathematics is a form of contemplation different from prayer or meditation, but for Jean-Pierre Changeux, the rigor and stringency of mathematics bring the human mind into contact with notions of God. A secular version of this is the belief that mathematics is the method for thinking explicitly about infinity, for making it apprehensible. Hermann Weyl says that "the goal of mathematics is the symbolic comprehension of the infinite with human, that is finite, means." This makes the most precise of sciences a way of approaching the most unknowable part of our existence.

To understand how past human beings had simplified ideas of infinity, it helps to remember that consciousness is an evolving project. Translations modernize texts, so we read ancient writing and think that it was done by people who appear to talk like we do, although sometimes more formally, and forget that they were often spirit thinkers who knew practically zero about how the world actually ran. Nearly every explanation they had for what they saw and experienced was mystical to the point of being irrational. Gods speaking through charms and chance, through entrails and the patterns of birds in flight, and lest one remark that such people were pagans and that Christianity corrected such thinking and made the world modern I would remind us—and it is no observation of my own—that the metaphor governing Christian Communion is a type of blood sacrifice.

We cannot know how earlier people felt intimations of a divine presence, because things are simply too different now. We have accounts from those figures in the desert that are vehement and awestruck. They had visions and visitations and experiences that aren't easy to fit into ordinary life, at least not our version of ordinary life. Words go only so far to acquaint us with a fierce and compelling experience had by someone so different from ourselves. And who knows what it was to live in that kind of darkness, which was so deep and enfolding as to have almost a texture. With no terrestrial light except firelight to compete with them, the stars must have looked as if they were hung just out of reach and to be as bright as neon. I wonder if the stars didn't seem to be as if hovering, observing, impersonal and imperial, the eyes of gods. The Greeks saw them as sufficiently animated as to find arrangements among them and to give these patterns names and qualities and purposes. Hysteria as a clinical circumstance more or less disappeared with candlelight's being replaced by gas and electric light, and who knows how many ancient authorities would these days be given a diagnosis and a prescription.

When I look at the night sky, I tend to think of it as the ancients did, as a place. The notion that the sky goes on forever is a modern one.

21.

Infinity occasionally contradicts intuition. There are as many fractions as there are whole numbers, for example, but there are also an infinite number of fractions between any two whole numbers. Between any two fractions, no matter how tightly bound, there is sufficient room to place an infinite number of other fractions.

An infinite collection of numbers exists, but there is also an infinite collection of numbers that haven't been imagined yet. As for geometry, in *Two New Sciences* Galileo noted the discordance in considering two line segments of unequal length. Each segment consists of an infinite number of points, but the longer segment would seem to have more points. He concluded that *less*, *greater*, and *equal* didn't apply to infinite things, and that there might also be a greater circumstance than infinity.

Something is infinite when it includes collections that

have as many terms as it does, Bertrand Russell writes. Something is also infinite when you can take things from it without making it smaller. Such observations, while concisely Western, also seem Zen-like.

Since the nineteenth century, dividing by zero has been regarded as undefined, contradictory, and illogical. Dividing 10 by smaller numbers, say 5 or 2, provides a clean answer. Dividing 10 by 7 or 3 is more complicated. Nevertheless, the larger dividing number provides the smaller answer. The inverse is also true; the smaller dividing number returns a bigger answer: 10 divided by 1/4 is 40. The answer to 10, or any other number's, being divided by zero would appear to be infinity.

An infinity that can be placed into a correspondence with the natural numbers is called countable, because each member of the infinite set can be assigned a counting number. The set of even numbers is countable, and so is the set of prime numbers.

1	2	3	4	5	. . .	
2	4	6	8	10	. . .	
2	3	5	7	11	. . .	

All three sets are the same size. An example of an uncountable infinity is the set of all the numbers there are—negative and positive, fractions, decimals such as pi that never end, and so on. These are called the real numbers. The most populous real numbers are the irrational numbers, which are numbers such as pi that can be written as a decimal but not as a fraction. There are so many irrational numbers that by the laws of probability if you pick a random number on the number line the odds that you picked an irrational number are 100%.

A list of real numbers will always be incomplete, because there is always a number that is not on the list. This is easily demonstrated: changing the digit corresponding to each number's place on the list produces an absent number. In other words, if your list begins .1234 . . . and continues .1235, ...1236, . . . and so on, you change the 1 in the first number to another number, you change the 2 in the second number, the 3 in the third, until, having changed all the endless digits of all the endless numbers, you will have a number that hasn't been included on your original list, verifying that the list of real numbers is too large to correspond to the list of whole numbers.

Pairing whole numbers and fractions toward the end of the nineteenth century, the German mathematician Georg Cantor, the founder of set theory, showed that there were as many whole numbers as fractions, even though there are an infinite number of fractions between two whole numbers. He also showed that there are more real numbers than fractions. Trying to pair off fractions with real numbers would always overlook a number, such as pi, that has an endless sequence of digits that don't repeat.

The Greeks thought of infinity concretely. Euclid wrote of prime numbers not that there are an infinite number of them but that they exceed "any assigned multitude of prime numbers." Until Cantor identified specific infinities, likely before 1878, no one had thought that there was more than one or such a thing as a higher infinity. Philosophers regarded infinity as either endlessly extensive or endlessly diminutive. Religious-minded people believed that being infinite was a characteristic of the divine, and mathematicians saw infinity as a boundary in calculus.

Before Cantor it had seemed self-evident that if you have a collection of things and you add things to it, the collection grows larger, and that taking things from it makes it smaller. Cantor realized that for each infinity, though, there would be a larger one, that there are a multitude of infinities, and that infinities ought to be able to be added to one another.

Infinity for Cantor consisted of collections, a many that could be thought of as a one, he said. Numbers that define the size of sets are called cardinal numbers. The cardinal number for a set of five objects is five. The cardinal number for a set of infinite objects is a transfinite cardinal, which is a conceptual object. The set of even numbers and the set of whole numbers, being the same size, have the same cardinality.

Set theory was also capable of constructing numbers and therefore infinities. Zero is the empty set. One is the set whose only member is the empty set. Two is the set containing the empty set and the set that contains the empty set. Three contains the empty set, the set containing the empty set, in the set made up of the empty set and the set containing the empty set.

Cantor was a Lutheran and, like Newton and Kepler, he believed that his work exemplified thoughts in the mind of God. Sometimes he heard a voice telling him that his conclusions were correct, no matter what anyone else said. In 1884, when he was thirty-nine, he was hospitalized for manic depression. Of his illness and confinement, he wrote in a letter that it had "in no way broken me, but in fact made me stronger inwardly, happier and more expectantly joyful." He also said that it had allowed him to think and work without being disturbed. For periods afterward, though, he withdrew from mathematics, and for a while he taught philosophy. He also became engaged by the belief that Francis Bacon had written Shakespeare's plays, a field now called Baconian theory.

It interests me that someone who struggled with stability created, with set theory, a field that might be regarded as permanently stable.

22.

Beginner algebra insists that a student be comfortable with symbols. Geometry, on the other hand, is visual. You can draw a triangle with a right angle and illustrate the Pythagorean theorem to a child. Temple Grandin, who says that she thinks in pictures, has written, "My area of weakness is in algebra, because there is no way to visualize it."

Kid algebra, I felt, required that I think. Starter geometry asked that I reason. The difference, as I see it, is that thinking is open-ended. Given the terms of an algebra problem, I had to decide which among them would help identify the variable the problem was considering. Each step in geometry requires that I make a judgment based on logical premises, whereas in algebra each step after the first step involves a single reply to the step before it. In simple algebra one thinks at the start of a problem, then enacts a procedure; in geometry one reasons to the end.

More than algebra, geometry engaged whatever intellectual resources I have. Algebra felt cloistered. Geometry

seemed expansive. Observing the saw-toothed profile of
the city skyline or the pattern of winter tree limbs like
lines on a grid made me feel that I lived among it. This
was obscurely cheering. I began to think that things actually are, in their essence, numbers, or that numbers are
at least capable of describing identities precisely. Algebra
has a mystic quality, while geometry at first appears prosaic, even though all mathematics, all pure mathematics
anyway, eventually engages the mystical. When I'm hitting on all cylinders, though, algebra seems procedural in
a way that geometry doesn't seem to be.

PERHAPS BECAUSE GEOMETRY relied on words used plainly,
a standard established in *Elements*, it seemed coherent in a
way that algebra is not. I did not feel adrift in geometry
among barely apprehensible and concealed possibilities,
as I did in algebra. In Euclidean geometry one intends
to prove an assertion and begins with the simplest thing
one can prove or assume from other proofs. In algebra,
having chosen a tactic, one pursues it to the end, where
the answer is correct or not. In geometry there is a self-
consciousness in that one takes a step then contemplates
the next one. There is not always a single correct answer.
The Pythagorean theorem has more than 317 proofs.
Leonardo da Vinci had one, James Garfield had one when
he was a congressman from Ohio, before he was president, and Einstein had one when he was twelve.

I PLANNED TO learn geometry from Euclid, which is how
I imagined that the pros did it. Also, I wanted to do it like
Abraham Lincoln had. I saw geometry as a return from an

exile amid the trials of algebra in order to spend a sojourn among the Greeks and pure thought. I imagined people asking what I was up to and answering, "At the moment? Reading Euclid," then basking in their admiration. For a while I carried *Elements* on the subway, like a badge.

Elements, it turns out, is very challenging. It almost hurt to read it. The propositions I could mostly understand, but the proofs were taxing. Proposition 2 is "From a given point to draw a straight line equal to a given straight line," which seems simple, draw a line equal to the length of another, but one arrives at the following drawing, after approximately twenty steps.

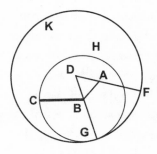

Grasping a proof involved a state of attention so laborious to enact that I grew impatient. I would give up and try again later and get nowhere one more time. It was the first occasion when I experienced mental work so demanding that sometimes I would feel that I had to lie down, but I wasn't actually tired. Except for *Finnegans Wake*, which I have never finished, nothing I had read was so difficult. It made me feel like a dunce. If I lived in ancient Greece, I would like to think that I would seek out Socrates or Plato or Aristotle, but I might cross the street to avoid Euclid. Don't ask a question unless you *really* want to know the answer, is how I might feel about him.

Amie told me that *Elements* hadn't been part of the reading for any class she'd taken in college or in graduate school and that she hadn't ever read it, and this made me feel free to seek other means of instruction.

23.

I found a geometry text and set to work, and a few weeks in I took a break and went to the Museum of Modern Art, where I stopped in front of *Broadway Boogie Woogie*, which was painted in 1943 by Piet Mondrian. The painting is sufficiently familiar that I probably don't need to say that it is a collection of grid-like yellow and red and blue and black rectangles and squares on a white ground, arranged up and down and across, like a crossword puzzle. The musician Jason Moran has said that it looks to him like a jazz score, and he has played it. From my novice engagement with geometry, the painting looked different to me from how it had before, more nervous, more brilliant as an observation, and more precise, as if a schematic depiction of movement in the city and in its way Pythagorean. A conceptual companion to Giacometti's love of watching people cross open spaces in the city, while noting the paths they chose and the patterns they formed as they moved among one another.

It would seem to serve no purpose to notice this, except that such an activity is a kind of theorizing about the world, a means of connecting one moment to another. And why is it necessary anyway that every thought have a purpose? Thoughts that appear to be peripheral lead sometimes toward thoughts that aren't. They recall and

suggest other thoughts and form ideas that become, partly
by aggregation, consequential. Everything that happens is
happening on the way to something else. This is a truism,
of course, but geometry made it apparent to me.

EVERY NOW AND then something in geometry seems familiar, a definition or the stages of a proof, and I realize,
I was here before. I remember then the other boys in my
class, and my teacher, whose name was Dawes Potter. He
was tall and in middle age, with an oval face and black
hair in a crew cut, and he always wore a bow tie, which
gave him an old-fashioned appearance. He looked like a
character in a comic strip. In my mind's eye, I can see him
observing me, and as an older man I know that the expression he is fixing on me is part confusion and part exasperation from not knowing what to do with a boy who
appeared otherwise to be intelligent, but sat sullen and
silent in class and much of the time simply stared out the
window. How obtuse he wouldn't even have known, since
my answers on paper were often the answers of the boy I
was sitting next to, a kind of academic ventriloquism.

24.

Set theory, divinity, and contradiction: for any set, there
is always a larger set—the set to which another element
has been added. Eventually one arrives at the set that
contains all other sets, "the single, completely individual
unity in which everything is included," Cantor wrote to
a friend in 1908. The set that contained all of its subsets,

however, would be larger than the set of everything, but anything cannot be larger than everything.

Cantor believed in his work "because I have followed its roots, so to speak, to the first infallible cause of all created things," and he resolved the paradox theologically. He wrote the mathematician Grace Chisholm Young that what surpasses all was "the single, completely individual unity in which everything is included, which includes the Absolute, incomprehensible to the human understanding. This is the *Actus Purissimus*, which by many is called God."

Since God is unknowable and absolute it made sense to Cantor that any finite system would collapse into paradox as one came near to the divine. Furthermore, in the region of absolute infinity it also made sense that the terms of the spatiotemporal realm would no longer hold. As Rilke writes in *Worpswede*, his monograph on landscape painting, "We shall often have to call a halt before the unknown."

A philosopher or theologian might have accepted Cantor, because there are always paradoxes and unresolved problems in dealing with deeply difficult questions, but mathematicians did not. Cantor was sketchy, and mathematics wasn't. "You were getting reasoning that seemed very simple," Gregory Chaitin told Lawrence Kuhn about Cantor, "and then you were saying 1 equals 0."

For some Christian theologians, though, God could be the only infinite. Cantor's inquiries as to the extent of an infinity were pointless, since they could have no answer.

BERTRAND RUSSELL FOUND a second paradox in set theory. In *Introduction to Mathematical Philosophy*, he suggests forming a class (or set) of all classes that are not members of themselves. "This is a class: is it a member of itself or not? If it is, it is one of those classes that are not members of themselves, i.e., it is not a member of itself. If it is not, it is not one of those classes that are not members of themselves, i.e. it is a member of itself. Thus of the two hypotheses—that it is, and that it is not, a member of itself—each implies its contradictory."

The discordance was addressed in the early twentieth century by the Zermelo-Fraenkel axioms of arithmetic, which are named for Ernst Zermelo and Abraham Fraenkel. The axioms provide a formal system that stipulates how sets are formed, and they do not allow a set of everything to be made.

25.

Very few ancient ideas regarding physical reality have any permanence, but the idea that numbers are the means for comprehending the design of the natural world persists. Pythagoras is supposed to have said, "Number is the measure of all things." Also, "Number rules the universe," and "Number is the within of all things." The adaptability of numbers allowed them to become a symbolic language that describes physical structures and movements. In more than one place I have read that absent numbers, physics is mute.

Mark Balaguer, a philosopher of mathematics, makes a distinction between high school math, which is a skill, and the math that lies beyond it, which is a theoretical practice, being "a theory about the world in the way that physics and biology are." In physics and biology, precise things are known about real objects, whereas in mathematics things are known infallibly about unreal objects. Both a scientist and a mathematician explore a territory, a habitat. The home ground of a species whose existence is conjectured tends to be circumscribed; the more capacious the habitat, the more likely the object is to have been found already. The territory, however, is real. A mathematician can bring a mathematical object into being only by thinking about it, and its home ground is abstract and infinite. In the essay "Mathematical Creation," published in 1910, Henri Poincaré writes that mathematics is "the activity in which the human mind seems to take least from the outside world."

MATHEMATICS IS SOMETIMES described as the study of structures, with the primary structure being 1, 2, 3, . . . Observations about this structure allowed the proposal of more elaborate structures. There is no explanation for why conclusions reached by an imaginary pursuit should describe processes and objects in the physical world, but occasionally they do. "An enigma presents itself which in all ages has agitated inquiring minds," Einstein said in Berlin in 1921 in an address to the Prussian Academy of Sciences. He asked how mathematics, "which is independent of experience, is so admirably appropriate to the objects of reality?"

Some of the patterns that numbers embody—the way they describe distance or temperature, for example—are simple and self-imposed, but mathematics sometimes also describes structures that suggest further structures. For a physicist the most desirable mathematics describes an observation and predicts others that then are found. "One seeks the most general ideas of operation which will bring together in simple, logical and unified form the largest possible circle of formal relationships," Einstein writes. This happened with Newton's law of gravitation, which described why objects fall to the ground and then was discovered also to describe the attractions among heavenly bodies and their orbits and also the tides.

The physicist Paul Dirac said, in 1939, "Pure mathematics and physics are becoming ever more closely connected, though their methods remain different. One may describe the situation by saying that the mathematician plays a game in which he himself invents the rules while the physicist plays a game in which the rules are provided by Nature, but as time goes on it becomes increasingly evident that the rules which the mathematician finds interesting are the same as those which Nature has chosen."

Like Pythagoras, Galileo saw mathematics as a text or a key to the physical world. "Philosophy is written in this grand book, the universe, which stands continually open to our gaze," he writes. "But the book cannot be understood unless one first learns to comprehend the language and read the characters in which it is written. It is written in the language of mathematics, and its characters are triangles, circles, and other geometric figures without which it is humanly impossible to understand a single word of it; without these one is wandering in a dark labyrinth." Such thinking makes mathematics seem to be, whatever else it is, a sacred code.

Many scientists believe as Pythagoras does that numbers are embodied in the natural world. That mathematics could make predictions about nature without performing experiments has no explanation, but "must be ascribed to some mathematical quality in Nature," Dirac says, "a quality which the casual observer of Nature would not suspect, but which nevertheless plays an important role in Nature's scheme."

In 1960 the physicist Eugene Wigner published an essay, "The Unreasonable Effectiveness of Mathematics in the Natural Sciences," which includes the observation that "the enormous usefulness of mathematics in the natural sciences is something bordering on the mysterious." Wigner wasn't surprised that mathematics described orbits, since he accepted that mathematics characterized the physical world. Wigner was surprised when a piece of pure mathematics, presumed to be totally abstract, was discovered to have a useful application, a circumstance that had no explanation.

The world gives an appearance of conforming to certain mathematical structures, and this idea is hard to shed entirely. The physicist Paul Davies told Lawrence Kuhn that the universe seems to be "a package of marvels. It's ingenious, it's ordered, it's mathematical." If there is no mathematical structure in nature, why does mathematics seem to find one? Davies has said that there might be a "primal accord between mathematics and the universe. The more that is discovered, the more this relation seems to be enforced."

C. N. Yang, a theoretical physicist, said, "What could be more mysterious, what could be more awe-inspiring, than to find that the structure of the physical world is intimately tied to the deep mathematical concepts, concepts which were developed out of considerations rooted only in logic and the beauty of form?" These of course are Platonic-shaded remarks.

In back of such a thought also is the appeal of sym-

metry. Symmetry and proportion have figured in Western culture at least since Plato, but they also appear in the
art and architecture of cultures that never heard of Plato.
Symmetry seems innately sympathetic to some quality
of being human. In *Metaphysics* Aristotle says, "The main
species of beauty are orderly arrangement, proportion,
and definiteness; and these are especially manifested by
the mathematical sciences." Mathematics evinces beauty
and goodness, he says, "in the highest degree." Such thinking might regard Newton as having uncovered, not imposed, a unification and to believe that further ones are to
be found. Since mathematics is the language of physics it
would suggest that any universal theory of physics would
also be a mathematical one.

26.

Three responses to the applicability of mathematics to
the physical world are common; the first two are com-
patible. One is that Pythagoras and Plato and Kepler and
Galileo and Einstein and Dirac are correct in assuming
that mathematicians and physicists find mathematical
formulas underlying the world's design because the world
is, inexplicably, built according to mathematical principles. The second is that mathematics is the most precise
language we have for describing physical structures and
so it is sensible that it would describe the physical world
better than anything else does and now and then with a
precision that is uncanny.

The third possibility is discordant with the others. It is
that human beings contribute an arrangement to nature

that isn't actually there. They do it without seeing that they do it. In "Cognitive Science and the Connection Between Physics and Mathematics," Anshu Gupta Mujumdar and Tejinder Singh write that mathematics results from prehistoric perceptions of "object, size, shape, pattern and change" from which have been built, by means of metaphor, "numbers, point, line," and the entire concrete and conceptual tackle box of mathematics. In physics, these concepts lead to ideas about force and movement and notions such as "field and symmetry." Since physics is built on the same ancient principles that brought mathematics into being, Mujumdar and Singh write, "this demystifies the extraordinary success of mathematics in physics. It could not have been otherwise." The correspondence between physics and mathematics is the result of the mind's being disposed to respond to regularity. Such a connection is "not mysterious. Rather, it is inevitable."

As with all categorical but (so far) unprovable claims, what can one reply except, "Possibly." And to remind oneself that the patterns that humans contribute to nature are nevertheless there.

Stanislas Dehaene also believes that mathematics has evolved to fit the structure of the physical world. "The evolution of mathematics is a fact," he writes. "Science historians have recorded its slow rise, through trial and error, to greater efficiency. It may not be necessary, then, to postulate that the universe was designed to conform to mathematical laws. Isn't it rather our mathematical laws, and the organizing principles of our brain before them, that were selected according to how closely they fit the structure of the universe? The miracle of the effectiveness of mathematics, dear to Eugene Wigner, could then be accounted for by selective evolution."

Neurology aside, is it possible that mathematics is more inclined to refinement than evolution? Evolution depends on accident, whereas mathematics depends on thinking to invoke it, and only in the sense of its being a creative and contemplative act does it include accident. Evolution is opportunistic. An adaptation succeeds because it works better than previous versions, whereas a piece of mathematics prevails because it is true. Epiphanies and coincidences, such as Newton's with the apple, are not the same as accidents of thought. Also, unlike as in evolution, earlier versions of mathematics are not discarded, they remain useful. Their primacy may lapse, but they don't become extinct. Except in the fossil record, you cannot find all the iterations of a species, but all the iterations of a piece of mathematics are extant. They don't get eliminated, they crowd up; in the guise of fields and specialties, they form neighborhoods.

There is not a trail of discredited and discarded mathematics or an archive of them, either, because mathematics is proved. Older mathematics might reflect a period's intellectual shortcomings, and sometimes it is wrong, but it doesn't tend to be wrong, it isn't *usually* wrong. There is no natural selection in mathematics, either. More powerful methods may ask earlier ones to step aside, but earlier ones occasionally return with uses not seen earlier. Sometimes areas of study with unsolved problems are eclipsed or are overlooked and return. Or, with applied mathematics, are solved by scientists in other fields. Or are put aside for being too difficult, or simply fall out of fashion. When I look for examples of this, I find things I don't entirely understand, such as a case given by Branko Grünbaum in his *Lectures on Lost Mathematics*, from 1975. Grünbaum says that there are problems "rooted in phenomena that are of interest to architects, engineers, some modern sculptors, and geometers: The rigidity or mobility of variously hinged systems of polygons, rods (= segments), cables, etc. Cauchy's theorem on the rigidity of polyhedra that have as faces rigid polygons hinged along common edges is probably the deepest known result." Cauchy's theorem is from the nineteenth century.

When I asked Amie for an example of a piece of mathematics that has fallen into disuse but is still useful, she wrote that she had asked her friends and one had said, "Cayley-Salmon theorem on the flecnode polynomial. This is a polynomial whose vanishing on a 2 dimensional surface in C^3 guarantees that the surface is ruled. (The theorem is that the surface is ruled if there is a line tangent to it to 3rd order at each point.)" Cayley's theorem on the flecnode polynomial appears in a book Cayley published in 1865. A librarian at the University of Chicago told Benson Farb, Amie's husband, that she regularly retrieves books and papers from earlier than the 1840s for mathematicians. Earlier mathematics typically forms part of a foundation.

Dehaene has other sensible objections to Wigner, though. Mathematicians have produced a gigantic amount of mathematics, he says, and only a small portion of it works in physics. A reply might be that pure mathematicians do not do mathematics in order to be useful to physicists, although if they did perhaps more mathematics might apply. Dehaene also observes that Kepler was not strictly correct in describing the orbits of the planets as ellipses. "The earth would perhaps follow an exact elliptic trajectory if it were alone in the solar system, if it was a perfect sphere, if it did not exchange energy with the sun, and so on," Dehaene writes. "In practice, however, all planets follow chaotic trajectories that merely resemble ellipses and are impossible to calculate precisely beyond a limit of several thousand years." This is completely true.

We (maybe just me) are still left with an (apparent) conundrum: If numbers and mathematical objects are mental constructions arising from the adaptation of the human brain to the regularities of the universe, then why are there primes, whose natures and behaviors have nothing to do with us? Also, it is still possible that, rather than an adaptation, the applicability of mathematics involves the observation of something that might be true and seems to hold throughout creation. We don't know for sure either way.

The mind sees what it is capable of seeing, Dehaene says, and scientists tend to overlook that "the brain is not a logical, universal, and optimal machine." It reasons poorly, he says, and does not do well with long chains of calculation (Amen). Also, it is subject to bias. Onto the physical world it projects structure "where only evolution and randomness are at work. Is the universe really 'written in mathematical language,' as Galileo contended? I am inclined to think instead that this is the only language with which we can try to read it."

On the other hand, Roger Penrose, a Nobel laureate, says, "It is undoubtedly the case that the more deeply we probe Nature's secrets, the more profoundly we are driven into Plato's world of mathematical ideals as we seek our understanding."

There is no settling the argument about Platonism or the relevance of numbers as describers of the natural world, not empirically anyway. There is only taking a side or being disinterested. The concept of a divine entity may be as far as mathematicians can go in their assertions before something powerful, enigmatic, and obdurate resists them, and they turn back, as if enacting Cantor's belief that as one approaches the far ranges of infinity the claims and assumptions of the spatiotemporal world break down. Dante's description of the Rose of Paradise is exquisite, but past it he isn't able to see clearly.

27.

I still don't totally know why starter math was so hard for me. I have wondered whether there are capacities native to all of us, the proportions of which determine what becomes congenial to our thinking. Did I lack ones favorable to mathematics? I have wondered whether there is a difference between procedural thinking and associative thinking. Certainly some people are capable of both and some, such as myself, can manage only degrees of one or the other. I think that the resistance I feel to studying math, which is nearly visceral, is a resistance to being made to think in a way that contradicts my nature or the way I am made or whatever you want to call it, a problem with authority, with being compelled to do something that allows no deviation. In the nature versus nurture argument I believe mainly in nature. We do not all begin at the same neurological starting line. Talent, I am persuaded, is a consequence of the way that our brains are built, and it is individual, either native or not. The circumstances into which we fall are surely important, and whom we fall among, but I'm not sure that we can exceed

temperamentally or psychically or perhaps even intellectually what we have it in us to become any more than, to
repeat a familiar observation, a young man or woman by
means of determination and without the given attributes
of size and strength can make it to the professional ranks
of basketball. The brain can compensate for gross senso-
rial deficits such as the loss of one's hearing or eyesight,
it is to some degree plastic, but not sufficiently to address
the neurological peculiarities involved with, say, autism.
This may be the consequence of the brain's being less
able to detect the subtler difficulties as deficits, although
it is also not my experience as the father of an autistic
child that all autistic people regard their capabilities as deficient, even though they are frequently told that they are.

So far as I can tell, what I also appear to be averse to
is solving problems that require holding several procedures in mind at once, the sort of thinking where one gets
three or more steps removed from the start, as happens
with equations. Circumstances where I have to identify
a variable, say, conceive of a formula or a function that
handles it, and complete the calculations while bearing
in mind the rules imposed by the formula or function.
Such procedures become something like automatic for
many people, so that such people are not always aware
even that they are thinking, but I have never found such
blithe competency simple to enact. Perhaps I spend too
much effort engaged with the procedures and not enough
on thinking broadly. Here, though, is where the capacity
for associative thinking figures. If I am reading a difficult
piece of writing, I don't try to understand it by noting the
steps in the writer's reasoning or design, although this is a
very good method. I allow myself, and this happens without my conscious engagement, to welcome associations,

many of which are dismissed until the ones that are left fall into place, assuming I am not defeated.

In neophyte math the problems are practical and literal but the solutions involve the apprehension of concepts. Going through them I feel like a plodder, one of the ungifted, the unblessed, blighted and meanwhile often aware that in the presence of difficult matters my mind might surrender. At times this can make me feel doubtful to the point of uneasiness about who I think I am and what my abilities are.

Some, often very intelligent, people design themselves around an aggressive embrace of their limitations, and often they are very successful. It makes their lives simpler than the confrontation with their deeper identities would entail. I see this division, for example, between people who believe that the world separates itself into types, and those who believe that each human life is singular. I fall adamantly into the second class, but I am aware that people in the first class arrive at judgments more quickly and tend to have more of them, although the judgments risk being shallow and ordinary.

Such people evaluate their observations according to a system. Would math be simpler for me if I had a systematic way of thinking? If I considered only categories and not individual cases? It's difficult to answer, of course, but I think, Probably. I feel at a loss when learning things that have an inherent order, and I feel unhappy at lacking the capacity. I have to focus on my ability to read the world differently, but, until the higher stages, which I will never reach, math doesn't reward individuality. Plus, as a result of this trait, I waste a lot of time.

It has made me wonder, was my inability a missing

talent or the consequence of having inhibited myself years ago, when math first became hopeless. There is no way for me to know, of course. I can't be, cognitively, two people at once, although, like anyone, I can be two and perhaps more people at once emotionally or psychically. I am trying to untangle this, since I am suggesting that emotional responses might be involved, an inhibition brought about by a surrender to being overwhelmed. Some, maybe many, of the behaviors we adopt in childhood are designed to protect us from harm, but they aren't always useful later in life. They don't always age well.

There is a further circumstance that my renewed engagement with mathematics insists I consider, also not attractive, which is that I might simply be lazy. Maybe I have avoided things that are hard to learn and to think about. Self-candor later in life has two sides. One side is the pleasure of correcting a fault, and the other is the reckoning of all the difficulty and harm that the fault has brought about, the gestures and episodes of defeat and chagrin included in the wagonload of small miseries that one is yoked to. The archive of mature experience more and more displays the capacity to be conflicted, or perhaps it is only that it grows richer. In the midst of celebrating, we are also mourning, which is an echo of the line from the Book of Common Prayer "In the midst of life we are in death," another observation that deepens for me with the years.

ONE CAN MEMORIZE, or one can aspire to understand. It's not always clear to the aspirant when one prevails over the other, or where the boundary between them is, either.

An actor speaking lines doesn't need to know why the writer wrote them. The knowledge might enlarge his or her performance, but it isn't required, I don't think. No one knows anything about the figures in Greek drama or what Shakespeare thought beyond what the plays and the poems suggest. Sometimes one has a cast of mind into which ideas land in their entirety, with their design revealed like a tool that arrives with its use understood intuitively. I don't think it is uncommon in mathematics for a young person to feel a reverberation with the forms that he or she is learning to use. Karen Olsson describes such a feeling, and I know that Amie has had it, too. Probably such a capacity makes learning math simpler, but it doesn't appear to be essential.

On this encounter I wanted to understand everything I could about the math I was learning; I even thought I would be able to, because I was older and the math I was studying was for kids, but this, I was forcefully made aware, was prideful. Simply too many elements are involved. The thought is too wide-ranging, which is not surprising, considering that even adolescent math is more than what Euclid knew. It was also borne in on me, something I might have realized earlier, that if I had the capacity to understand math now, I might not have had so much trouble in the first place.

Sometimes I wondered if maybe I had been taught poorly. Or by being inept and by means of embarrassment had I just fallen through the cracks? I'm not sure. I remember the child that I was, I can close my eyes and see him and hear him speak and watch him, but I am no longer that same being. He is independent of me. He exists in his own time.

28.

Mathematics draws on the human desire to simplify. To divine that what appears to be disorderly can be sorted and classified. One can see this impulse at work in procedures and taxonomies. The urge to classify seems as ancient as the beings who drew animals on the walls of caves. Whatever profound spirit thinking they were engaged in, they also were organizing categories, drawing herds, grouping like with like.

From a journal: Mathematics doesn't solicit my opinion. Math doesn't ask "Are you comfortable?" It is not congenial that way. Mathematics is cold, but there is a kind of liberation in its aloofness. It is an inverted liberation, though, a whose-service-is-perfect-freedom sort, since it comes about only by means of a submission to its terms.

What have I found difficult lately? I am baffled by the absurd idea that in an expression such as $2 \times 10^5 \, m^3 \times 1.1$ kg/m^3 I can shed the cubic meters, since they cancel each other, but cubic meters are *things* and not numbers. I understand that I can cancel numbers, which are abstract, but a meter is a noun and material. How is their equivalence possible?

$$2 \times 10^5 \; \cancel{m^3} \bullet 1.1 \, \frac{kg}{\cancel{m^3}}$$

If I write, 2×10^5 turtles$^3 \times 1.1$ kg/turtles^3, can I get rid of the turtles? As it turns out, yes, Amie says. "The turtles would be a unit in the problem you're solving," she writes. "Maybe each turtle eats 1.1 kg of turtle food and you'd like to figure out how much turtle food, in kilos, to order for 200,000 turtles, which is 2×10^5 turtles. Your final answer is in kilograms. The turtle units get canceled."

Another moment when mathematics seems to depend on a loopy, self-serving, and clandestine logic. I read of a mathematician's saying that sometimes solutions appear from unlikely places and procedures as by a kind of witchcraft, and I thought, Exactly.

29.

"Geometry is a completely idealistic world," Deane said, invoking Plato. "It's about things that don't really exist—the perfect circle, the perfect square. Things that are roughly a square or roughly a circle, they exist. Math exists only in your head, not anywhere else."

Plato said, "The knowledge of which geometry aims is the knowledge of the eternal." The remark, "He is unworthy of the name of man who is ignorant of the fact that the diagonal of a square is incommensurable with its side," is frequently attributed to him on the internet and said to have been written over the entrance to his academy, but if he said it, I can't find where, nor could the scholars I consulted.

30.

As everyone had said, geometry turned out to be more congenial than algebra had been. I wouldn't say it was welcoming, but I wasn't roughed up, either. I didn't come out of the encounter bruised and disheveled.

With geometry I had fewer days where I was exas-

perated. Algebra seemed to like to smack me around, because it could. What algebra I had possession of, though,
I could use in analytical geometry, which was discovered
by Descartes in the seventeenth century and involves
graphing lines and curves using algebraic equations. As
a result, I wasn't beginning geometry as the same rube I
had been when I opened *Algebra for Dummies*.

There continued to be periods, though, when I was serially defeated. I agree that defeat is not to be feared, but
too much defeat is. A defeat surplus is. Defeat as a condition is withering. The company of failure can seem as permanent as a tattoo. I was not cheered by failing so often. I
did not think, How great it will be when all this falls into
place. I worried that it would never fall into place. That
there was no place for it to fall into even. In math I am
failing against a pitiless standard. There is no ambiguity,
whereas failure in the eyes of others, as a writer, say, is a
matter of other people's judgment, of taste and bias and
opinion, all of which are culturally conditioned and can
be questioned. There are very few absolute standards to
life, however much a person might sometimes wish that
there were.

When people tell you how to deal with failure, they
usually mean in the singular. They mean, You failed,
here's what you do. They don't mean, You keep failing,
even though you are trying. Nothing is working for you.
Maybe give up, find something else. Perhaps they tell you
that things can change. Most of us would avoid someone
who seemed tagged with constant defeat. What came
next for me, though, day upon day, was more of what I'd
been failing at. I didn't have simply to get through something. Apparently I had to be smarter than I actually was.

Perseverance furthers, though, and I was determined

not to give up. Anyway, I could choose to view the daily failing cumulatively, as a single failure amounting to a judgment: I guess I'm not good at math.

To be unable to fulfill an intellectual task is frustrating, though. It is like being found to be weak, which no one likes, either. I turned away from math as an adolescent because it dented my hopes of becoming a capable person. It is vexing to find that it still has the capacity to undermine me.

Of course, math is actually taxing in a demonstrable way. In *The Number Sense* Stanislas Dehaene mentions that brain scans show that subtracting 3 repeatedly involves "intense bilateral activation of parietal and frontal lobes. If an operation as elementary as subtraction already mobilizes our neuronal network to such an extent, one can imagine the concentration and the level of expertise needed to demonstrate a novel and truly difficult mathematical conjecture! It is not so surprising, then, that error and imprecision so often mar mathematical constructions."

I found this cheering. Everyone needs to make his or her own mistakes, and one of mine might have been to study math. We have things to tell each other in the form of advice and instruction, but very little of it is ironclad or will mean the same to one person as another. Experience is individual. And making a mistake is maybe not the same thing as failing.

Even so, failing is daunting. Research by Judith Rodin at Yale suggests that mastering a task later in life makes a person more confident and that this confidence can be used to attempt other, presumably more demanding, tasks. I couldn't find any research on the subject of failure later in life. It's a territory one would rather avoid.

.   .   .

AT TIMES WITH geometry the annoyance I had felt with algebra returned. Sometimes I'd get angry, and then I did stupid things repetitively. It would begin by my noting that a solution, say for finding the proportions between line segments on a graph, had varied perversely from what (I believed) I had been led to think was the correct way to find them. I would get a problem wrong and attempt another, insisting on the same approach, as if I could force the problem to surrender, as if I could break its spirit.

On such days, I rebuked Amie for perpetuating the lie that math was rational. Clearly when you interrogated it, contradictions fell out of it all over the place. I never got anywhere with this argument, though. Math in her mind was blameless.

31.

The shapes in geometry are straightforward and simple, but Euclid examines them and their attributes in such a way that they aren't simple anymore. The ubiquity of lines, circles, squares, and rectangles in the natural world makes you feel as if you were seeing something fundamental about its design. One can imagine the Greeks' astonishment at the world's seeming to reflect their suppositions. Or at believing that they had discovered something essential lurking just below its surface, seeing in the flight of a bird a parabola and regarding it as a confirmation.

A deep engagement with pure mathematics is like prayer, in that one is attempting to speak to an unknown power, across a divide, and awaiting an answer.

32.

Applied mathematics, a case: Poker played poorly is purely a gambler's game. Losers tend to think that they didn't get the cards, and not that they were beaten by someone who played better than they did. They return to the table and wait for big hands and lose more. Accomplished players strive to diminish the effects of luck. From their opponents' bets and behaviors, they work like detectives to determine their cards. They play opportune hands deceptively, and feckless ones, too, and shed unpromising ones before the cards cause them too much harm. They know that some hands that seem auspicious are not, and that others are stronger than they appear.

Games for which a flawless strategy is known are said to be solved. Tic-tac-toe is solved; blackjack is solved; checkers is solved. Chess is not solved, and poker is not, either. Solutions theoretically exist; they are simply too intricate, so far, to be comprehended. Among mathematicians, chess is regarded as a game of perfect information, because nothing is hidden. If its ideal strategy were discovered, there would no longer be any reason to play it—no move could be made for which the response was not already known. Poker is a game of imperfect information, since so much is concealed. Solving it would not overcome the disadvantage of being unable to know why your opponent is acting as he or she is. Such concepts derive from an abstruse field of applied mathematics called game theory, which was formulated, in the 1940s, to address difficult economic problems.

Game theory was conceptualized by John von Neumann, who was one of the mathematicians involved in the Manhattan Project, and who collaborated with Einstein. In 1944, von Neumann, with Oskar Morgenstern, published *Theory of Games and Economic Behavior*. Until then, people typically entered markets with a strategy, but such preparation could help them only if they knew what other people would do. Von Neumann saw that the tidal rhythms of transactions and uncertainties involved in markets were embodied in the narratives of parlor games, and especially in poker, where each player also has a strategy to claim the largest share of the money changing hands. Since the strategies of most games were subject, if simplified, to fairly concise mathematical calculations, the workings of markets could be also.

Theory of Games and Economic Behavior is riddled with charts, equations, and diagrams. Without an understanding of higher math, it is impenetrable. In an essay reprinted in the book's sixtieth-anniversary edition, John McDonald, quoting John Maynard Keynes, writes, "'Businessmen play a mixed game of skill and chance, the average results of which to the players are not known by those who take a hand.' Von Neumann's theory is designed to narrow this gamble . . . It tries to make the imponderable ponderable."

To diagram certain game theory problems, von Neumann used hands of poker as examples. Fifty years later, Chris Ferguson, a UCLA undergraduate, thought to apply game theory concepts to grand-master poker. Relying on them, he became, in 2000, the first person to win more than a million dollars in a poker tournament.

FERGUSON IS TALL and lanky, with very long brown hair and a brown goatee; his admirers call him Jesus. When he plays cards, he wears a black cowboy hat that he was given by a friend. The brim protrudes over his forehead like an overhang, sometimes throwing his eyes, behind dark glasses, into shadow. At the poker table, he has all the animation of a state trooper handing out a speeding ticket. When he's especially engaged, he sits almost cataleptically still, with his hands clasped in front of his chin. His right hand is balled into a fist, and his left hand rests open on top of it. To bet, he lowers his right arm like a lever, then returns to his original pose. The gesture is exactly the same whether the bet is a bluff or a boast.

One year at the World Series of Poker, in Las Vegas, Ferguson played poker ten hours a day for thirty-five days in a row. He had a room in the Rio Hotel, where the tournament was held, and when he was finished playing, usually around three or four in the morning, he went upstairs and slept until noon, then came back to the lobby and sat down at a table. A few times, he finished at seven in the morning. Eventually, he found himself unable to fall asleep any earlier than that.

I went to his room one night so that he could show me some fancy forms of shuffling—cheat forms, shuffles that allow dealers to deal specific cards. Leaning on one elbow, he used the bed for a table and showed me a double shuffle, a shuffle that appears to be genuine but isn't a shuffle at all—all the cards remain where they were. "I learned it for protection, in the nineties," he said, "to see if people were doing it to me. A really good magician is going to beat a good spotter. A good dealer is supposed to deal in a way that you know he can't be cheating."

Ferguson was born in Los Angeles in 1963. His mother, Beatriz, was a mathematician and his father, Tom, taught game theory at UCLA. Tom Ferguson brought home specialized board games and card games and taught them to Chris and his older brother, Marc, who is a computer programmer. "Whenever there was a rainy day, we would get to stay inside and play Risk," Tom Ferguson told me. His younger son "learned to think about playing and strategies and what other people know about what you know. It's not important in chess, but it's important in poker. It's a rather deep game, when you get involved."

Ferguson doesn't recall when his father taught him poker—he feels as if he's known it all his life—but he remembers that when he was in the fourth grade he lost thirty-five cents in a game, and it bothered him. In high school, he played on the weekend with friends. At seventeen, accompanied by some of them, he began making trips to Las Vegas. He and his friends would pool gas money, sleep in cheap hotels, and eat at the buffets. He liked Vegas because the people in the casinos called him Sir, and "you could lean back in your chair and no one would yell at you, unlike school." His friends went for fun, but Ferguson went to establish whether he played poker well. "I saw Vegas as a challenge," he told me. "Play ten hours a day, pay for my room and my food, and get home with more money than I started with. I think of that as my transition into manhood—when I was able to prove to myself that if I had no support from my family, and no job, as long as I could get to Vegas, and have a hundred dollars in my pocket, I could survive."

From a fifty-two-card deck, 2,598,960 five-card hands are possible. The basis for most poker strategy is a ruthless notion: What can I discern about my opponents' habits that I can attack? Such an approach is called "maximally exploitive." It is the way nearly all professionals proceed, relying on logic and intuition. While Ferguson was still a student, he decided also to employ a method called "optimal strategy," which derives from a game theory question posed by Claude Chevalley, in 1945, in *View*: "Each player being ignorant of the strategies followed by his opponents, which strategy will he follow in order to get the maximum possible advantage for himself?"

The optimal strategy "doesn't mean 'How do I win the most?'" Ferguson says. It means, when up against an expert opponent, "How do I lose the least?" Part of it is mathematically determining whether one's cards are favorable, but a player using optimal strategy also builds into his play bets that sometimes appear improbable and make it mathematically difficult for the opponent to know what to do. With optimal strategy, "if we're playing heads up, you might get lucky and beat me, but you'll never outplay me," Ferguson said.

Ferguson made trips to Las Vegas during his five years as an undergraduate and his thirteen years as a graduate student in computer science. On the entry forms of poker tournaments, for many years, he listed his occupation as "student." His thesis adviser was Leonard Kleinrock, whose lab sent what was considered the first message over the internet, in 1969. Kleinrock told me that Ferguson was "one of the more brilliant and creative young men that I've known in my career at UCLA." During the late eighties, Ferguson was working as a programmer for a more advanced doctoral candidate when the student got a result he couldn't interpret. "Chris, this lowly pro-grammer, writing code, explained what was causing the result," Kleinrock said. "It was a very deep theoretical idea, and his manner was very low-key, no bravado, just pure intelligence, and, when I saw that, I thought, I want to follow through with this guy."

Kleinrock said that Ferguson "would spend hours bouncing ideas around. All kinds of esoteric mathemat-ical and computer subjects—genetic algorithms, search algorithms, and so on—reams of ideas, then he'd come back the next day pursuing some of them, having thrown the others away." He went on, "Or he would show me how he was progressing on cutting the deck down to any number of cards—sixteenth card or thirty-fourth card—and the perfect riffle of them. I was not the type of super-visor to demand a schedule—we were both enjoying the academic and scientific ideas. He was, by far, the student who took the longest to graduate, though." Ferguson fin-ished his doctorate in 1999, when he was thirty-six, by which time he had spent half his life at UCLA. "He was never the go-go-go academic achiever that wanted to race up there and set the world on fire," Kleinrock said. "Plus, the year after he graduated he became the world cham-pion and won all that money."

33.

Poker allows two ways to win: own the best hand, or make the best hand go away, sometimes by bluffing. The imperative to bluff, it turns out, is inherent. "I can take a purely mathematical model of poker and hunt for

a purely mathematical solution," Harold Kuhn, a professor at Princeton who knew von Neumann, told me, "and
a phenomenon will appear which has always seemed to
be psychological but isn't—it's mathematical—which
is that you will bluff, and your opponent will drop out.
The necessity for bluffing is built into the mathematics of
the model."

Any hand has multiple strategies, Ferguson said. "If I'm
facing someone I've never played before and have no idea
of his weakness, and I play an optimal strategy, I know I
will not make a mistake that will give him any money,"
he said. "If you don't know the optimal strategy, you don't
know your weaknesses or his; you don't know when he's
taking advantage of you and you can take advantage of
him. As people deviate from optimal strategy—as they
bluff or fold or call too often or not enough—it's actually
pretty clear. If you're able to see how they deviate, you
can see how to take advantage of them."

A player using optimal strategy assumes that his opponents know he is doing so—in other words, that his strategy has been found out. He can announce, for example,
that a third of his bets will be bluffs, and then construct
the game in such a way that his opponent still can't tell
whether it is better to fold or call. If two players have each
put fifty dollars into the pot, and the optimal-strategy
player bets a hundred dollars and his opponent folds, the
opponent loses fifty dollars. If he calls, one-third of the
time he will win, because the optimal-strategy player is
bluffing, and two-thirds of the time he will lose, because
the optimal-strategy player is betting a hand that is strong
enough to win. The opponent now has no means of
knowing when it is better to call than to fold. This is de

scribed as making the opponent "indifferent." He might as well flip a coin. "Now it's a mind game," Ferguson said.

"What are the guys who don't play optimal strategy doing?" I asked.

"I'm not sure what they're thinking," Ferguson said. "They're flying by the seat of their pants. I learned poker by sitting at home and thinking how to play hands—if I play my hands this way, what can my opponent do to take advantage of me, and if he can, what do I need to do so that he can't anymore? I want to be the least exploitable player. Other people learn through experience, and if they're good they're going to come up with a strategy that's pretty similar to what I do. It turns out that there's just a right way to play. I learned by applying game theory. They learned through what I consider a more arduous process, playing countless hands. Am I smarter because I use game theory? I don't think so. It's hard to learn poker, because you can play a hand horribly and win, and also play perfectly—almost—and lose. How's the guy who doesn't know the game well going to know the difference?"

Some players think that approaching poker through mathematics causes someone like Ferguson to lose sight of all the peripheral elements of the game, such as "tells," unconsciously revealing behaviors that intuitive players regard as rich in information. Ferguson doesn't dismiss these aspects of the game—he both looks for tells and tries not to display them—but he believes that game theory protects him from making intuitive judgments that might fail, or from being distracted by information that's not necessarily germane. Exploitive players "model their opponents, and then they come up with a strategy that will beat that opponent—'He raised here, he must have that hand' or 'He plays that way, I'm going to play this way'—and they stop there," Ferguson said. "I don't stop there. I say, 'If I play this way, how can he play to counteract what I'm doing—how might he adjust?' The beauty of it is that it doesn't depend on your opponent. Once I figure out what the optimal strategy is, I know it. A year from now, it will be the same. It doesn't matter who I'm playing against. The research is everlasting."

Another player who uses game theory and mathematics as heavily as Ferguson does is Andy Bloch, who has two degrees in electrical engineering from MIT and a law degree from Harvard. "Most other people are trying to outplay you—bluff you out of pots, trying to get a read on you," Bloch told me. "If you're playing against someone like that, you can manipulate them into making bad calls and folds. Chris is one of the most difficult people to get an edge on, if you can get an edge at all. Against other players, I'm going to try, but with Chris I don't even really try. He's too difficult to read. Players unfamiliar with game theory, the intuitive players, are going to have a really hard time reading and understanding him, because some of the plays he makes are going to confuse them. They'll see a bluff and think he bluffs too much, because the bluff doesn't make sense."

34.

At the tail end of the World Series of Poker that year, the Bellagio held a tournament, and Ferguson entered, although he was exhausted—he had played in thirty-three of fifty-five events of the series, and had won around

seven hundred thousand dollars. The buy-in at the Bellagio was fifteen thousand dollars, cash only. Ferguson
paid with three five-thousand-dollar chips that a friend
had given him to satisfy a debt. He sat with eight other
players at a table in a high-ceilinged room with huge win-
dows at one end, off the casino proper. In all, there were
four hundred and forty-six players at forty-five tables.
There was a low, sibilant rustle in the room, the sound of
chips being agitated in hundreds of hands. On the walls
were television screens showing a golf tournament and
an action movie, with the sound turned off.

Ferguson calls himself a tournament specialist, meaning that he doesn't play for table stakes—what are known
as live games. In live games, players can always buy more
chips when they lose. Ferguson regards this as tedious. In
a tournament, when you've lost your chips, you're done. A
tournament player has to accumulate chips to withstand
challenges, which become more consequential as the
match progresses. Players are often allowed to enter tournament games a few hours after play has started, which is
what he likes to do. "I lose the advantage of knowing the
table," he says, meaning the other players, but, because he
has rested, his decisions are better.

Thousands of men and women are believed to play
poker for a living. The entry fees at big tournaments can
be so high that chairs are sometimes filled by people representing syndicates—groups of players who have pooled
their money and sent one player to represent them. If
there are winnings, they split them.

Ferguson kept getting cards he didn't want. He turned
their edges over slightly to read them, then tossed them
back and folded his arms across his chest and looked im-
passively at the movie. Now and then he leaned back on

the legs of his chair. He was roused briefly by a man in a white T-shirt and black shorts, with shiny black hair, who walked mostly sideways among the tables and chairs toward the door. "I'm a friggin' idiot," the man said. "That's what I am. Stone-cold idiot."

I stood near Ferguson's table. Julio Rodriguez, who writes for CardPlayer.com, was filing updates on the game. I asked Rodriguez how unusual Ferguson was in being so fluent in math. "A lot of players know the math," he said. "There's no way around it, really, but the majority of them go on instinct, or feel, or a read on a player. It seems like a lot of them are just born with a sense of games. If you talk to a lot of these guys outside, they're never not playing a game. They're reading you. That's why they're so engaging. They know what people want; it's very easy for them to please you. They also know how to deceive."

I asked why he thought Ferguson appeared to be playing so cautiously. This early in the tournament, he said, "everyone's mostly sitting on their hands, waiting for someone to make a mistake. Chris is very stoic. He'll wait as long as he has to. He's watching the other players."

"They're watching him, too?"

Rodriguez pursed his lips and shook his head slightly, as if to say the effort would be pointless. "He's like a robot," he said. "You get nothing."

As it happened, the afternoon went poorly for Ferguson. Almost two hours in, he had lost nearly half his chips. His expression remained flat. He lifted his hat a few times and scratched his head. Three times in a row he was raised out of a pot. Much of the time, his cards had been good enough to get him into the hand but not sufficient to allow him to stay. He called a ten-thousand-dollar bet and lost.

Finally, he shoved all his chips toward the center of the table, showed his cards, and was out. He shook hands with the other players and wished them well, reminding me of something his father had said: "One of the nice things about Chris growing up and playing games is he didn't mind losing. Most kids I play with are scared of losing. He knows that losing, you learn something." I asked Chris if losing bothered him, and he said it only really upset him if he lost as the result of a mistake. Anyway, he said, thinking you would win all the time at poker was unrealistic.

FERGUSON AND I went to dinner at a Mexican restaurant at the Palms. The restaurant didn't have a table ready. The hostess gave us a beeper, so that we could wander around the casino. Ferguson was tired and wanted to sit down. The first chair we found was in front of a row of video poker machines. I put a dollar in the machine.

"This is something I never do," Ferguson said. The machine offered several games. "The only one I know the optimal strategy for is Jacks or Better."

"Play that," I said.

"They don't have it," he said. He chose one called Super Double Double, the closest he could find to Jacks or Better, and five cards appeared on the screen. He played quickly, hitting a button to hold or to draw cards.

A pair of tens arrived. "This is a really bad hand," he said, and drew three more cards, none of which improved it. He talked as he played. "I'm trying to figure out whether to hold ace-queen, to the two fours, because the payout is different," he said. And, "Here's what I'm thinking: pair of queens and the sevens. I'm throwing away the sevens—I don't think there's any question about it." And, "What's the payoff to me if I just held the kings? There's a chance I get another king about one-eighth of the time. There's a chance I get four kings; I'm guessing that's an additional eighth unit toward the payout—I'm up to three-eighths. I need to get up to a half."

The red light on our beeper began flashing. Two nines appeared. Ferguson discarded the three other cards, and in the draw he got two more nines. At the cashier's cage, I collected fourteen dollars and applied it to the dinner.

35.

Mathematics being the same everywhere means that one does not have to study Arabic math or Chinese math or Canadian math. Math isn't balkanized. If you are of the correct temperament or talent or intellect, mathematics in its higher ranges is about the intellectual reach of humanity and the contents of your own thinking.

Sometimes I didn't understand what I was doing, but I could do it anyway. This was the case with radians, which, like degrees, are a means of measuring angles. A 360-degree angle is a 2pi radian angle. Figuring the ratios of circles according to their radians, I felt as I did the work, flawlessly as it turned out for a change, that I understood nothing of the concepts I was enacting. The task, I suppose, is not that difficult, but I almost never understood how I had reached the answer, except to say that the formulas had provided it.

Occasionally I would come to something that I remembered partially from childhood, such as tangents and their relationship to a circle, and I would have a sinking feeling, recalling how difficult the subject had been, but also wondering if I would understand it now and would I understand why I hadn't been able to learn it before. Having approached it more deliberately this time, I expected to do better. This wasn't always the case. At times, when problems got difficult, and I failed one after another, I felt as if I'd been left at a senior center and given a textbook of problems to pass my day, because the attendants had been told I enjoyed math. Something had happened to my powers of reasoning, though, and I couldn't solve the problems anymore.

I end up talking back to books. That's impossible, that's stupid. You're wrong. You made a mistake, idiot book.

IN *Mathematics: A Very Short Introduction*, by Timothy Gowers, I encounter the notion that to succeed at math "one should learn to think abstractly, because by doing so many philosophical difficulties simply disappear." I hadn't considered that besides the practical difficulties I was having, I was also having philosophical ones. Learning to become less rigid in my thinking with math was like submitting to the charm of the books I argued with. I did my best to submit to math's charm, but I didn't always think it was charming. Doing math sometimes felt like playing with a dog that bites.

THE COMPLICATED STRUCTURES of mathematics can be explained the way any other work of art can be explained, except that understanding mathematics requires more learning than most of us possess. I can understand the outline of what Amie does, sort of, but if I turn the pages of, say, "The Cohomological Equation for Partially Hyperbolic Diffeomorphisms," which is among her more cited papers, I cannot grasp any of its assertions. At best, I recognize a few symbols. Another of her papers is called "Stably Ergodic Approximation: Two Examples." I consistently misread this as "Sadly Ergodic Approximation," which sounds like the title of a piece of music by Frank Zappa.

Someone to whom a theorem has been explained, depending on its complexity, might feel like someone who has learned to play a complicated piece of music and now understands its design. He or she might appreciate the way certain passages are formed, the logic applied, the rules of harmony engaged or acted against. This is a satisfaction of art, of being made aware of deeper relationships than the ones that are plain on a first encounter. You have completed a task. You can play that piece of music now. You can try to play other pieces based on its logic and the technique you acquired in learning it. You have broadened yourself, your abilities, perhaps your capacity for wonder, expanded your life list. You have seen something of the design that the world suggests. Ever elusive. Ever hovering at the edges of sight, of apprehension. Such awareness is consoling. It finds for us a place in the world. A standpoint from which another horizon is always in view.

Summer

1.

When I finally reached calculus, I was somewhere I had never been. I could have said that algebra introduces abstractions into arithmetic, and geometry studies forms, but I couldn't have said what calculus does. I expected it to be beyond my capabilities and was gratified when only some of it was.

Calculus departs almost completely from arithmetic, by which I mean that numbers do not really explain it. Nor does the logic of geometry. The definition of calculus that I find most often is that calculus is a means of describing instantaneous change. In *A Tour of the Calculus*, David Berlinski writes that calculus measures "how far an object has been going fast," and "how fast an object has gone far."

Another description I have heard is that calculus uses linear forms to bring precision to nonlinear forms—straight lines to measure curved lines such as circles, ellipses, and parabolas, which depicted on a graph might represent, among other things, movement or distance or the rate at which something is increasing or decreasing or time, all of these being cases of change. A turning line on a graph can't be measured precisely, for example, but a straight line, technically a line segment, can. Likewise, a turning path can be approximated by dividing it into segments of straight lines. Adding the lengths of the segments describes the path. The smaller the segments, the more precisely they portray the curving line. Calculus then is a series of approximations that become more and more accurate as the measures are refined. Anything that changes demonstrably, both actual and imaginary things, including markets and machines and designs for potential machines, is within the range of calculus.

The measurements can never be exact, because of a concept called a limit. A limit is an endpoint in an investigation, a destination that you can get arbitrarily close to but usually never reach. In the case of a turning path, the limit is where you would arrive if you could take an infinite number of steps, which you can't. A limit is exemplified in Zeno's paradoxes, the most famous of which probably is the race between Achilles and the tortoise, in which the tortoise gets a head start. Zeno's argument is that Achilles can never pass the tortoise, because every place he reaches is somewhere that the tortoise has already left. He can get half as close, a quarter as close, and so on.

Approaching a limit, measurements cluster as finer and finer reductions are made. In measuring distance, consider the diminishing sequence 1, 1/2, 1/4, 1/8, 1/16, 1/32, . . . The figures do not reach their limit, which is zero distance, because eventually they would approach $1/\infty$, which is undefined, meaning that it does not exist, division by infinity being impossible. Even so, the sequence approaching zero has an infinite number of members.

If speed is being measured, the time considered cannot be reduced to zero, because at zero there is no movement. An object moves only between two moments.

A limit allows precise rather than average descriptions of change. Average speed is simply distance divided by time. Driving 60 miles over 2 hours means an average speed of 30 miles per hour. Calculus determines the speed over an interval—how fast the car was going between 48 and 49 miles, say. Between 48 miles and 48 miles and 5 feet. Between 48 miles and 48 miles and 6 inches and so on.

There are two kinds of calculus, differential and integral. Differential calculus divides up time—how fast an object has gone far (or how fast a process is accumulating). Integral divides up distance—how far an object has gone fast (or how much is accumulating). To assess the progress of a car trip, say, you could plot the trip on a graph on which the x axis is time, and the y axis is distance. The trip would begin where the axes intersect, at the point 0,0, and proceed as a line on the graph. On the x axis the trip's endpoint would represent how long the trip took, and on the y axis the endpoint would represent how far you had gone. (If you had ended where you started, left home and returned, perhaps, the endpoint on the x axis would be 0.)

The line would curve, because sometimes you sped up and sometimes you slowed down, and if you stopped it would be flat until you started again. To reveal speed, you find the slope of the line between two points—the difference, that is, between the heights at each point—and divide it by the time the line encloses; this is called rise over run, a phrase with its own kind of poetry and one of the few things in mathematics that I remembered easily.

As the distance between the points gets closer to zero, the slopes approach the derivative. A version of the derivative is what a speedometer provides.

Knowing how fast you'd been going, you can make a second graph where the x axis also represents time but the y axis represents speed instead of distance. The measurement of the area between two points on the line and their coordinates on the x axis tells how far you had traveled between any two moments. This is the integral. Finding the derivative uses mainly algebra, in the form of difference in ratios, and finding the integral uses geometry in the form of computing area. One operation reverses the other. The first case used distance as a function of time, and by means of differentiating—finding the derivative, that is—figured out how fast you were going. Measuring the area beneath the curve on the graph, by means of integration, revealed how far you had gone. By differentiating and then integrating, one returns to where one began. This is elegant and surprising and exemplifies the fundamental theorem of calculus, which says that differential and integral calculus are inversions of each other. This much about calculus I was able to absorb after much effort.

2.

I don't want to say that Amie was mistaken, but word problems return in student calculus. A car stops, a man or a woman gets out. The car drives away. How fast is it moving when it is twenty feet from him or her? A quarter mile? Ten miles and thirty-four inches? And how long has it taken to get there? I have also encountered calculus as the means for solving problems such as, What is the price at which an item will draw the largest number of buyers and make the most money? How long should you own a car before the depreciation and the repairs make it no longer sensible?

When I came to these, I did something I didn't do in algebra: I turned the page. I thought, Life is too short.

3.

Algebra and geometry are earthbound and practical, and there doesn't seem to be anything mysterious about their simpler versions. They have their origins in the immediately visible world. How to determine the boundaries of a piece of land or establish the characteristics of forms. Geometry is a bridge to abstract thinking in being bound up with suppositions about the world's design, but calculus is about being made intimate with unseen things, about making the difficult to observe comprehensible. Algebra and geometry *imply* time, in the unfolding of their procedures; calculus embodies time.

I found starter calculus formidable in a way that starter algebra and geometry aren't. Calculus contains powerful tools that are not so easy to understand, although (some of them) manageable by means of persistence. The pleasure of using calculus, though, compared with algebra and geometry, is greater even at the simple level I am acquainted with. One feels engaged with larger powers and occasionally capable of impressive gestures; at least they impressed me. I had as much difficulty with calculus as I had with algebra and geometry, maybe more, but I thought I was cool while I was doing it.

CALCULUS WAS CREATED to provide answers for problems in physics that beforehand could be done only painstakingly or by improvised methods. Isaac Newton is regarded as having discovered calculus, but Gottfried Leibniz worked on similar material and published before Newton, who for roughly ten years had kept his work to himself. Newton was in England, and Leibniz was in Germany, and while Newton is given more credit than Leibniz is, Leibniz's notation is the one that is used.

Newton organized disparate procedures into a formal order. He was intending to provide a working method, rather than prove the theoretical basis for one. In this sense, Deane says, it was as if he had begun in midair rather than on the ground the way someone such as Euclid did. Unlike Euclid, also, Newton wasn't specifically engaged in extending the range of mathematics. When his suppositions were correct, he pressed forward instead of working backward to establish a foundation, as another mathematician might have. Newton's influence, partly, was to favor intuition as a source for mathematical ideas, for finding problems in the physical world and inventing mathematics for solving them. Before him, most mathematics involved solving problems presented by earlier work. An attitude about pure mathematics begins with Newton.

Physicists tend to take on a concept that perhaps has an intuitive meaning and use it to develop a tool or a theory. When they see it describe something in nature, they figure it works. Mathematicians examine the concept and demonstrate that it actually has a rigorous basis in mathematical axioms. This is called axiomatizing physics.

Physics is descriptive and not generally concerned with logic. A physicist cares about calculations that describe and predict observed phenomena. There are no external phenomena in mathematics. There are only assertions and axioms and their consequences. Things are pursued, but they are abstract things, and one can never be completely sure that the sought-after thing will appear. There is also the possibility that the mathematics will refute it.

THE ELEGANCE, THE ambition, the sweep of calculus signify a moment when the world stilled and a spectacular truth came into being, although only for one person, or maybe two, to give Leibniz his due. Einstein said of calculus that it was "the greatest advance in thought that a single individual was ever privileged to make."

Since change is pervasive and everlasting, calculus applies widely; in mathematics it is a small part in a lot of fields, and in the real world it applies in engineering, commerce, markets, stock trading, statistics, and medicine, among many other fields. The things that calculus describes—the gradients of change in a world that fluctuates constantly and permanently at all degrees of its scale—are so intrinsic and essential to understanding movement and time that it seems like it would have been found sooner. The Greeks had possession of some of the means necessary. Archimedes's measuring the area of a circle by drawing straight lines is conceptually connected, but the Greeks conducted mathematics in the context of proofs, to which calculus didn't initially lend

itself, and furthermore, while the Greeks were aware of
irrational numbers, they didn't commonly use them or
for a while even accept them, and irrational numbers are
required in order to express fine degrees of measurement.
Maybe, also, the world changed less dramatically in front
of their eyes. The progress of a ship toward the horizon
or from the horizon to the harbor or the transits of the
sky and the heavens happened so gradually that perhaps
they didn't suggest a subject for investigation.

4.

Newton's law of universal gravitation describes a mathematical pattern that is real but invisible. The circumstances resemble the abstract quality of numbers and
mathematical objects in that you can see them only when
they are represented by something else, when they are
practical metaphors (AAAA). Pure mathematics must
be invoked to be seen, but Newton described something
ubiquitous that is both concrete and hidden.

In *A Tour of the Calculus* Berlinski writes that with
Newton one has the impression of math's being used
to explore parts of the world that are beyond what the
senses explain. With *The Principia*, the proper title of
which is *Philosophiæ Naturalis Principia Mathematica*,
Newton ends the version of the universe as being without governing principles. Before Newton, the planets and
the stars had been assumed to behave according to laws
known by the spirits or gods or God, depending on the period. There were celestial laws for the heavens and terrestrial laws for earth, and they were assumed to be different

from each other. The Pythagorean and Euclidean notions of the world's being built from mathematical forms were speculative and mystical, whereas Newton established that the portrayal of the physical world by mathematics is actual. Berlinski describes the universe after Newton as being "coordinated by a Great Plan, an elaborate and densely reticulated set of mathematical laws."

For Christmas Day 1942, the tercentenary of Newton's birth, the Royal Society of London planned a celebration, but the war prevented it. The economist John Maynard Keynes was to have addressed the ceremony, but he died a few months before it was finally held, in 1946, so his brother read the remarks he had written. Newton was widely regarded as a modern figure, the first figure of the Enlightenment, but he was also a transitional one, being, Keynes said, "the last of the magicians, the last of the Babylonians and Sumerians, the last great mind which looked out on the visible and intellectual world with the same eyes as those who began to build our intellectual inheritance rather less than ten thousand years ago."

Newton thought of the universe and its contents as "a secret which could be read by applying pure thought to certain evidence, certain mystic clues which God had laid about the world to allow a sort of philosopher's treasure hunt," Keynes said. (These are remarks that might also be applied to Kepler.)

5.

I thought that if I could learn calculus I would begin to apprehend that behind ordinary appearances were patterns that were stable but also constantly enacting themselves. Or perhaps not behind but more closely at hand, so as to see relations that unfolded over discrete intervals, something at times like the rolling patterns of waves in deep water.

I am right up against melodrama here, but I needed romantic assumptions to sustain me sometimes, especially as the work got harder.

6.

Nearly a year in, after it was too late to do much about it, I realized that I had become separated from ordinary life. My wife would leave for work, dressed differently each day, and I wore the same clothes all week. There was something solemn and monk-like about my confinement. For hours I turned the pages of books, and everything took place in my head. I might as well have been working by candlelight. It wore on my spirits to be removed from the rest of the world. I had immersed myself in math, and I hadn't expected to. I had thought it would be easier to learn. I thought it would be a laudable and high-minded pastime, but math became all that I did and thought about. I had an abstruse hobby, not a hobby, an obsession. Everyone was interested when it came up, but no one wanted to hear about it once they saw that I'd gone a little crackers with it. They were mostly math averse, and I struck them as zealous. I could listen when people spoke to me, I could nod and appear to be paying attention, but I wasn't really. I had no idea how long some of my silences were.

7.

The attractive equation $\int_a^b f'(x)dx = f(b) - f(a)$ says that the integral from a to b of f' is equal to $f(b)$ minus $f(a)$—f of b, that is, minus f of a. It took me much more time than it should have to understand that the notation dx stands for change in x and that d is not a number to be multiplied by x, as it would be in algebra. It is a fundamental distinction, and trying to work the problem without understanding it is hopeless, but no one had explained it to me, and I had failed to grasp its meaning in whatever texts I'd encountered it in, which was unfortunate. I lost a painful amount of time trying to figure out why my answers were not only wrong but wildly wrong. Exotically wrong, comically wrong, I might have thought, if I had any tolerance left for being wrong.

dx is not an instruction to perform a task. It's a phrase, or maybe an abbreviation, for change with respect to x. Also, x times dx is not dx^2, but xdx. dx in algebra treats d as a number; xdx in calculus is a concept, not a procedure. To try to work out why none of this was clear to me, I turned to *Calculus Made Simple*, by H. Mulholland, and only became more deeply lost and also indignant at the title. Simple for who, I'd like to know. In *Mathematics for the Nonmathematician*, a non-mathematician appears to be someone with at least a college-level familiarity with, if not a higher degree in, mathematics. I tried *Calculus Made Easy*, by Silvanus P. Thompson, which was published in 1910, partly on the assumption that I ought to be at least as smart as people were a hundred years ago. On page 20, Thompson says that he is employing the binomial theorem, which, if I need to be reminded of it, he says is on page 116. I do need to be reminded of it. Actually, I need to be introduced to it.

Putting aside that I haven't ever read a book where the writer tells me that the information I need appears much later in the book, I make the trip to page 116, where I find:

Accordingly, we will avail ourselves of the binomial theorem, and expand the expression $\left(1 + \dfrac{1}{n}\right)^n$ in that well-known way.

The binomial theorem gives the rule that

$$(a + b)^n = a^n + n\frac{a^{n-1}b}{1!} + n(n - 1)\frac{a^{n-2}b^2}{2!}$$
$$+ n(n - 1)(n - 2)\frac{a^{n-3}b^3}{3!} + \text{etc.}$$

Since Thompson died in 1916, I can't ask him what he meant by "well-known way." I'd like to rebuke him retroactively, though. Something valuable shouldn't by ineptness be made more difficult to learn. That should be unacceptable. It awoke my resentment of haughty explainers who can't explain things clearly.

At the worst moments with calculus, I try to write out what I am attempting to solve, as I had been able to do sometimes with algebra, and I find that often I can't even frame the question. There are also periods when I spend my time unwisely. Accidentally dropping elements of equations, trying to get perfect scores, but making arithmetical mistakes, or losing track of positive and negative signs, which is my specialty and easier than you might think. You start at the beginning with algebra and geometry, but calculus is a compendium. Each time you nod your head in calculus, you are affirming three or four principles. It is a dense exchange.

Mathematics has a quality like a spiritual practice in

that I can't pretend to know what I don't know. I can guess
at an answer, as I had in high school, and of course sometimes I will be right but mostly I won't. Math has taught
me in its own severe way that I can't trust the passing
view, the glance. The world being as perilous as mathematics is, it's helpful to be reminded that I need to attend
the present, while also considering the past—that I need
to pay attention. One consequential judgment made or action taken without the support of thought and reason is
hazardous in life, as in math. The most reliable way for me
to leave the trail of the correct answer is to take the figuring too fast or cavalierly. Even in an empty room, at the
edge of old age, I feel the pressure of the classroom, and
the requirement from my childhood to be the bright boy.

MATH AND LANGUAGE may be connected in the minds of
some linguists, but learning math does not seem to me to
be like learning another language. It is like learning a language when you don't have one yet. There are rules and
circumstances to grasp before you even guess that you
can use this tool to express thoughts and feelings. I may
be inclined this way, though, since I am doing this late
and my learning is compressed. When we are children,
the span of time between learning to count and learning
algebra is long, and there are years to achieve some sort
of facility with numbers before we are expected to use
them in complicated ways.

ON A ROUGH day with calculus, opening a textbook, I see
a blizzard of numbers and symbols, and think of other
things that I might rather be doing. Somewhere late in

algebra, or maybe early in geometry, is when I realized that new knowledge lingers fitfully in the mind, and that it isn't possible to remember everything I am learning. An acquaintance is what I am managing. What I wish is that in high school, while studying geometry, say, I had also been taught the ways in which it figures in painting and perspective, in architecture and in the natural world and so on, so that I could see it perhaps the way the Greeks had seen it. If I had felt that the world was connected in its parts, I might have been provoked to a kind of wonder and enthusiasm. I might have *wanted* to learn. In approaching mathematics now through its separate disciplines, I wonder if I have misaligned my purposes. Perhaps I should have done what Deane suggested and looked more for mathematics in the world and less on the page.

What I might have seen as a child if I had paid closer attention or perhaps been more receptive is that behind the lessons I was learning was a form of instruction that would equip me to think for myself, a necessary condition of life. An advantage of this endeavor has been learning to think differently. It has been partly a project of renovation.

8.

To return to my parallel theme: believing in Platonism, at least in some degree, means accepting a companion reality that can't be seen or located or have its dimensions made plain, and, except to say that it is entirely other, can't be described beyond generalizations. Like numbers, human experience has abstract and practical sides, feeling and reason, and in each of us one or the other tends to dominate. Belief does not persuade a scientist, and science does not persuade a believer. Too ardent an embrace of reason leads to irrational thinking, and too ardent an embrace of feeling leads to madness. William James says that religious mysticism is only half of the possible mysticisms, the others are forms of insanity. These are the states in which mystical convictions circle back on a person and pessimistically invert notions of divinity into notions of evil.

Feeling responds to intuition, which arrives unbidden and suddenly. The suddenness by itself often persuades, Bertrand Russell says, but reason decides whether an intuition is reliable. In mathematics, as in art, intuition produces thoughts that reason enacts.

The boundary between these states is fluid and encloses the territory occupied by mysticism, of which there are two types. One type involves believing that there is more to life than we apprehend. The other includes people who have had a mystical experience. For as long as I can remember, I have been among the first type. If a mystical experience is one that makes a person aware of being a smaller part of a large design, then I guess I belong to the second type, too, if maybe weakly.

I'm not proud of this, especially. Having mystical inclinations seems a little simple. A means of thinking about the world that isn't too taxing or penetrating, but is satisfying and makes one feel a connection to deeper forces and broad in one's outlook, and, in a vain way, select, with finer capacities than many people have. It isn't rigorous, though, and maybe it is even shallow. Mystics sometimes appear to be insufficiently tough-minded and indifferent to reason, to be credulous and unsophisticated versions of theists. Mysticism has the appeal of the express arrival at an end, rather than working one's way patiently toward a position. We say, I *know* this is true, but what we mean is, I *feel* it to be true, or I believe it to be true, or, secretly, I wish it were true.

On the other hand, perhaps mysticism is an attempt to find a larger arrangement, a place where myth and science intersect, the boundary between the inner and outer lives. "The most beautiful thing we can experience is the mysterious," Einstein said. "It is the source of all true art and science. He to whom the emotion is a stranger, who can no longer pause to wonder and stand wrapped in awe, is as good as dead."

9.

The companion region in which Platonism resides has no significant divisions. According to Russell, all things being one, there is no past or future. The world, as Leonard Meyer describes it in *Music, the Arts, and Ideas*, is "a complex, continuous, single event." This is perhaps an Existential mysticism as much as a Platonic one.*

---

* Years ago when I was an LSD dabbler I had the experience, lying on my back on a mesa, that all moments were one moment and that time was a continuum with no beginning or end. I thought, This must be what holy men know. Then I heard a swooshing, aerodynamic sound, and I saw that a vulture had homed in on me, and I stood up and waved my arms and shouted, "I'm not dead." Even now, though, I wonder,

In *Naming Infinity*, Loren Graham and Jean-Michel Kantor write that mathematics has been the means of approaching the Absolute in all ages and cultures from "the classical Greek period, to the pre-Socratics, the Egyptians and the Babylonians," and in "China, India, and in Muslim, Jewish, Christian, and Buddhist tradition." Another means of considering mystical inclinations is to regard what we see as the surface of something else, as the outskirts of a further reality, areas of which mathematics describes. This can suggest the inner life and the psyche or the larger one represented by the cosmos. Animism, astrology, spirit thinking, the Greek and Roman gods, gods of all kinds and races are descriptions of an engagement with the ineffable. "The insight into the mystery of life, coupled though it be with fear, has also given rise to religion," Einstein says. "To know what is impenetrable to us really exists, manifesting itself as the highest wisdom and the most radiant beauty, which our dull faculties can comprehend only in their most primitive forms—this knowledge, this feeling is at the center of true religiousness."

I WOULD PREFER not to regard my receptivity to mysticism as a flaw. I could try to rid myself of it, to grow up, as it were, but this seems like trying to learn to throw with one's opposite hand. And what would be the point. Try-

---

are all moments somehow one moment, or had a change in the chemistry in my brain simply persuaded me that they were? Had I briefly visited an exalted and actual state of being, one adjacent to ordinary awareness, or did I imagine that I had? Had I perhaps had an intimation of the non-spatiotemporal realm? I find I can't answer the question. I raise it only to point out that in some circumstances what is the inner world and what is the outer world and how they are joined is not completely clear.

ing not to think about something leads to thinking about it more, in my experience. I would rather believe that it reminds me that human life contains layers of consequence and meaning, and can be contemplated endlessly without necessarily arriving at an end, a point of view that is itself mystical; apparently this type of thinking is circular. Perhaps, so far as thinking is concerned, there is no end to arrive at, or we are not yet anyway equipped to arrive at one.

Russell says that mysticism is better embraced as an attitude than a creed. "By sufficient restraint, there is an element of wisdom to be learned from the mystical way of feeling, which does not seem to be attainable in any other manner," he writes. Overblown mysticism is a misled product of human feeling. As an element among thought and feeling, though, mysticism "is the inspirer of whatever is best in Man," Russell says. "Even the cautious and patient investigation of truth by science, which seems the very antithesis of the mystic's swift certainty, may be fostered and nourished by that very spirit of reverence in which mysticism lives and moves." The Russian mathematician Nikolai Luzin wrote that while there are disciplines such as physics and biology that rely on the senses for insight, and logic and mathematics, which rely on reasoning, there was a third means of understanding, generally overlooked by scholars, which was "intuitive-mystical understanding."

By temperament I am more inclined to observe than to appraise. I haven't really got any critical faculties. I'm an enthusiast, a species of surveyor, the sort of person who cannot pass a hole in the ground without stopping to look into it.

10.

A religious mystic believes that God is understood better through illumination than through proposition and appraisal. An example of a non-mystic religious thinker would be Descartes, who thought that he could prove God's existence by logical means.

Some people who use mathematics to try to find God are irrationally excited, but there are also great mathematicians who feel that mathematics exemplifies something elementary and elusive about existence. Pythagoras, Blaise Pascal, Georg Cantor, and Hermann Weyl all had mystical inclinations. In *God and the Universe*, Weyl writes that mathematics "lifts the human mind into closer proximity with the divine than is attainable through any other medium." Pascal believed that the two infinities, small and large, were enigmas offered by nature that, as Alain Connes observes, could be admired but not understood. We are poised between the two infinities surrounding either end of our lives the way mysticism is poised between reason and feeling.

I believe the mystical realm exists, but this is a declaration of faith. Julian Jaynes proposes in *The Origin of Consciousness in the Breakdown of the Bicameral Mind* that human consciousness as we know it develops in the period between *The Iliad* and *The Odyssey*. In *The Iliad* the instructing voices come from elsewhere, from fogs and hallucinated figures, he says, whereas in *The Odyssey*, they reside at least partly in the minds of the characters. The impulses behind creative thought have never been easy to locate. An artist doesn't always know the sources of his or her work. Perhaps the images were created in their minds, but perhaps artists have a capacity or a receptivity that is greater than in more conventional sensibilities or perhaps they resist it less than others do. I cannot say that this view is not sentimental. I prefer to think that there is more to life than we apprehend, but clearly it isn't something I can prove. Arguing the existence of something that is immaterial is difficult, even though people try to all the time.

IN *The Varieties of Religious Experience*, William James writes that someone who has had a mystical experience can convey its quality only vaguely and indirectly. Moreover, someone not receptive to obscure states of being cannot appreciate what one actually is and is likely to think that the person describing one is being ridiculous.

Mystic states "are states of insight into depths of truth unplumbed by the discursive intellect," James writes. They appear to be transitory and transcendent positions between thought and feeling. Moreover, they have a successive effect in that ones that return seem to deepen the meanings of earlier ones. Finally, they are independent of conscious control. They might be facilitated by preparations "as by fixing the attention, or going through certain bodily performances, or in other ways which manuals of mysticism prescribe," but basically they arrive when they arrive.

The capacity to entertain mystical feelings perhaps exists in all of us but is elevated only in some of us. Whatever it is, "We are alive or dead to the eternal inner message of the arts according as we have kept or lost this mystical susceptibility," James writes. We are separated from mystical experiences "by the filmiest of screens," he continues. "We may go through life without suspecting their existence; but apply the requisite stimulus, and at a touch they are there in all their completeness, definite types of mentality which probably somewhere have their field of application and adaptation. No account of the universe in its totality can be final which leaves these other forms of consciousness quite disregarded. How to regard them is the question." James quotes a man receptive to such experiences who wondered if they were the same as what the saints describe and are not "the undemonstrable but irrefragable certainty of God?"

Mystical states, like occasions of faith, are "absolutely authoritative over the individuals to whom they come," James writes, but they don't often persuade anyone else. They demonstrate, though, that there is more to consciousness than the form that we accept as dominant. Kantor and Graham believe that the ineffable states that James describes arise during mathematical work, "moments when mathematicians refer to 'marvelous intuitions.'" In mathematics, according to Plato, "one seems to dream of essence."

I wonder how in our times a God-evoking experience might be obtained, but I think it would also be a terrifying one. It would be difficult not to think that one was going crazy. The description of God that I find most persuasive is the one by the southern evangelist who told the writer Philip Hamburger that God was a "luminous, oblong blur."

11.

With calculus I learned that I had to do even more than I had become accustomed to doing. I had to teach myself to

follow an explanation to its end instead of seeking flaws
as it was unfolding or looking for exits. I had to suppress
my anxieties.

I haven't encountered a border with mathematics that
is precisely defined, but if I go farther, and maybe not all
that much farther, I will find material I simply can't comprehend even if it's explained to me patiently. That, I suppose, is where I will have to stop.

12.

Artists can quote other artists, but only for effect. They
can quote straightforwardly or stealthily or ironically.
Mathematicians quote each other constantly, by applying
theorems and methods, but always in full and without
irony.

Controversy exists in mathematics, but it is not general or common. When it arises it tends to be specific
to a case—is this or that piece of work new, or was it
borrowed from someone else? Even when something is
discovered to be borrowed, it is not overthrown, the way,
appropriation art aside, a piece of plagiarism in the other
arts, even if accidental, might be; it is merely given its
proper attribution. In *The Weil Conjectures*, Karen Olsson
describes how André Weil published a result that was
called the Weil conjecture, although not the ones he published in 1949, which are still attached to his name. When
it became known that Weil's work had drawn heavily on
conversations with Yutaka Taniyama and Goro Shimura,
the name was changed to the Taniyama-Weil conjecture,
then the Taniyama-Shimura-Weil conjecture, and now

commonly, Olsson writes, to "the modularity conjecture for elliptic curves."

Reading and rereading formulas and equations and theorems in the attempt to understand them is different from looking for meaning in difficult prose or poetry. A literary reader is seeking a writer's intentions, which may be willfully concealed, or possibly not even clear to the writer. Mathematics prohibits obscurity, at least of a willful kind. In mathematics, the clearer something is, the more powerful it is, and the more useful. To obfuscate in math restricts the reach of the idea involved. In writing, in speaking, obfuscation has its place as a means of refinement, concealment, ornamentation, and even prestige. To write prose or poetry that is difficult to understand, to restrict intentionally one's audience, is to elevate the importance of one's work in the minds of some people. Sometimes, of course, it is merely pretension, a form of incompetence.

I can read a sentence over and over in prose or poetry and not be sure I understand what the writer intends, but I can continue reading, knowing usually that my uncertainty will not mean that I can't later find something larger and deeper in the design. So far as I can tell, you can't advance in math without understanding the design and its references. You don't always have to have the end in sight, but you have to see a way to it or part of a way. You have to know where to begin. I don't know if my mind doesn't find this form of thinking sympathetic, or if it is simply a matter of the material's being difficult. Still, it is a strange thing to confront something that I can almost but not quite understand, something that I know is not ambiguous, that has a single, universally understood meaning for those who grasp it, a remark that is a statement of fact, a logical extension of something I have been understanding, following, even if imperfectly, but am now lost with. An increase of thinking is required, and I don't know why sometimes I can't make it. Is there such a thing as thinking everything that one can, terminal velocity thinking? Or a state where no more thoughts are possible until some thoughts are shed? Once it became clear that Amie wouldn't be able to steer me through all my difficulties, perhaps I should have engaged a tutor, but that's against the rules that I had set for myself. It would be as if I had determined to build a house and was calling in a carpenter for the parts that were hard or seemed to be beyond my capacities. If I did that, I wouldn't be able to regard my house as my own work. This is pedantic, but those are the terms I adopted.

13.

In shedding "Math is inconsistent," for the awareness that the fault is usually mine, I exchange a confidence in my own judgment for a more vulnerable position. On the other hand, I am not unhappy to be reminded that it is better to ask a question than to assume an answer. Humility requires a denial of self, of a portion of self anyway, the flamboyant, the insecure, the part that fears being overlooked at the party, a phrase I am borrowing from William Maxwell. To allow another person, to allow everyone, the same regard, the same justification, is an achievement of psychic life, and a lot works within us to prevent it. And anyway, just coming up with a good question represents an advance. Cantor said that to ask

the right question is sometimes more difficult than to answer it.

Humility has been forced on me by my engaging in a
pursuit that I appear to be unfitted for. I had expected as
I neared the end to feel pleased at succeeding where I'd
once failed. To have enjoyed a small triumph in setting a
private record straight, and I have, a little, but not to the
degree I'd hoped for. If I had turned out to be capable at
math, I suppose I could have blamed other people for my
failing, but blaming others for one's poor performance is
never attractive.

14.

The days of difficulty accrue. As much as Amie tries to
help, the problem of someone's explaining something
to you, especially something complex, is that they don't
know what you don't know, or why it is hard for you
to know it. To be taught a difficult subject intimately is
sometimes to wonder how human beings manage to talk
to each other at all.

Calculus occasionally became so challenging that I
would think, Why do I need to know this? Why am I persisting in trying to learn a discipline that not a single ordinary civilian I know has possession of and that is of no
use to me anyway? The questions were beside the point.
I don't need to know calculus. I need to make the as-if-
pilgrimage of learning it as well as I can manage, because
it asks that I practice patience and discipline and think
clearly and try hard. I need to attempt a task that appears
to be beyond my abilities and that insists that I think in

ways I am not accustomed to thinking or feel comfortable attempting. Learning calculus also embodies my intuition, or perhaps only my desire, that dormant within
me are capacities that might be enlivened and make me
more than I am. I am hoping to meet a better version
of myself that I haven't yet become.

It seems important to continue enlarging ourselves
for as long as we can and to claim new territory for our
thinking. We are all given talents, and I don't want to arrive at the end feeling that I have not made what I might
have of my own, however modest they might be. From
the lives that I have witnessed the ends of, I know that
feeling disappointed by what one has made of oneself is
no way to prepare for a decent finish.

15.

Rules in calculus about how to handle formulas and functions, such as one called the chain rule, fall in a territory
between numbers and words, making them difficult for
me. While being aware that math has an organic quality,
that the materials I am studying have developed from one
another, in no place more apparent, at least in these early
stages, than in Euclid and in calculus, I nevertheless have
difficulty discerning the designs. When I was lost among
textbooks, I would sometimes watch calculus videos on
the Khan Academy website, and not infrequently I would
feel like I was observing a magic trick and thinking, How
did that happen?

Calculus exemplifies patterns but not always, or at
least that is how it seems to me. A function, for example,

is a sort of machine into which you enter values, and it delivers a single response. $f(x) = x^2$ is a simple function that delivers squares. An x you enter on the left-hand side of the equal sign becomes its square on the right-hand side. The variations in functions are endless. You can have a simple function, $f(x) = x^2 + 1$. You can have a function $f(x) = x^2 + 1$ if x is even, and $x^2 + 2$ if x is odd or the other way around. Equations that have more than one solution cannot serve as functions: $x^2 + y^2 = 8$, for example, cannot be a function, because if you solve for y by entering a number for x, say, 2, you get $4 + y^2 = 8$, which becomes $y^2 = 4$, which becomes $y = 2, -2$, since, while the square root of any positive integer is positive, it also has a negative duplicate, a companion on the other side of the looking glass, as it were.

Certain functions are named for their characteristics. $f(x) = x^2$ is a square function. $f(x) = x^3$ is a cube function. $f(x) = \sqrt{x}$ is a square root function. $f(x) = 1/x$ is a reciprocal function. Using f to denote a function is common but not a rule. Functions can also be combined. A combined function is called a composite function, which in beginner calculus is often written as $f(g(x))$. This means perform on x whatever operation g involves and enter that result into the function that f represents. The chain rule finds the derivatives of composite functions; that is, it finds the rates at which they are changing, it differentiates them. It says that the derivative of $f(g(x))$ is $f'(g(x)) \cdot g'(x)$. The designation f', which is read as f prime, signifies the derivative of f, and g' is the result of differentiating g. (The function $\sin(x^2)$ is the composition of $f(x) = \sin(x)$ and $g(x) = x^2$.) $f'(g(x))$ means take the derivative of f and evaluate it on a graph at the point $g(x)$. The derivative measures the slope of the tangent line on a graph at whatever point is chosen, and represents the instantaneous rate of change, meaning the limit of lines that are nearly tangent to the point.

This is not especially difficult, but it is not easy, either. The simplest way I can think of to put it is to say that to differentiate a composite function, you identify the functions that compose it and apply the chain rule. I am aware that the transaction appears to involve invoking a clear-cut procedure, but it was not clear-cut for me. The only reason I raise it is to discuss my next difficulty, which was the case of integration.

In identifying the integral, that is, the accumulation, from the derivative using the fundamental theorem of calculus, you have to figure out what the integral is derived from—that is, you have your answer, but you have to determine what provided it. Amie says differentiating is like following a recipe, say, for pancakes, whereas integrating is having the pancakes and trying to determine what they are made from. A process called u substitution undoes the chain rule, by working backward, and I could follow it, sort of, but only in the simplest cases. U substitution is where I hit the rocks hard.

U substitution involves the notation du/dx, which looks like a fraction but isn't. It means the derivative of u with respect to x, but there are times when you can *pretend* it's a fraction, if that happens to make your problem easier to solve. Exactly when you can do that is something I don't even now understand.

"We don't think backwards naturally," Amie said, trying to talk me down from the ledge. "We *do* think backwards naturally with murder mysteries, for example, we work backwards to the conclusion, but we don't do that in math until we get to calculus. Math up to that point is applying a set of rules, and you hit integration, and you have to apply guesses. You have to deduce things indirectly and understand objects by the way they behave, not by what they are." When she intuited how lost I was, she added, sympathetically, "It's not well defined."

Deane said, "The problem with integration is that there is no systematic way to do it, like there is with differentiation. Integration involves trying to guess which techniques might work, then doing the calculations and seeing if they provide a path to a solution. The first step, guessing what might work, can be challenging."

I about quit when he said that.

ALL I COULD do was go over it again and again, in the hope that familiarity, even if merely in the form of a blunt determination, would cause it to reveal itself. I was a little consoled to find that my circumstances were not singular. In *A Tour of the Calculus*, David Berlinski writes, "The eye slows; a feeling of helplessness steals over the soul. At first, it seems as if the confident language of mathematical assertion constitutes a subtle form of mockery. There is no help for any of this save the ancient remedies of practice and a willingness to put pencil to paper."

WHILE BEING SENTENCES, equations carry emphasis and meaning differently from the way that words do. Prose and poetry stimulate the memory, provoke layers of associations or form them, whereas an equation is freighted with rules that one must have command of as a means of translation. Formulas, equations, and functions are commands, not suggestions. In prose and poetry the literal meaning is sometimes a placeholder for other meanings.

An equation, while also a placeholder, is typically a vessel for a single thought. It might include many thoughts in terms of its composition and reach, but it delivers a single meaning. The references, the associations, are, by requirement, particular and without ambiguity. It is, in that sense, an ideal form. The rules allow your memories to be attached only to these references and in circumscribed ways. Its instructions are explicit and unbending.

If your mathematical vocabulary is narrow and shallow, as mine is, you might struggle to manage the information a formula or an equation contains or to understand how to use it. Sometimes one has to perform a type of translation. For example, seeing a statement such as $(\ln(\sin x))^3$, I had to remind myself that *ln* stands for the natural logarithm, something that, like u substitution, undoes a procedure, the raising of a number, called the base, to an exponent. $10^3 = 1000$. The logarithm of 1000 with regard to a base of 10 is 3. The natural logarithm is something different. The natural logarithm involves a base with a number called *e*, which is also called Euler's number and is approximately 2.718. $\ln(e^3) = 3$.

Then I had also to know that there are two procedures within what appears to be one: $\ln(\sin x)^3$, which is to say, the natural log taken of the sin of *x*, and the natural log of $\sin(x)$ then taken to the third power. To differentiate this expression I perform the chain rule on it, which finds the elements that compose the expression and multiplies their derivatives by each other. I have to understand what $\sin x$ means, what a logarithm is and what a natural log is, what its derivative is, and what it means to find the derivative of the sin of the term represented by *x* and then to multiply it by the other term, not to mention using the power rule to find the derivative and also to know what the exponent means. Years of learning are involved, and it's a simple destination, encountered during the first year of calculus, but a pilgrim has to have been on the road, undergoing trials, for a while to get there.

16.

I felt occasionally like a figure in a myth or a folk tale, attempting a scary or impossible task and having to compose myself in the face of doubts. More than in algebra and geometry, in calculus it seems as if a number of answers might apply, there appear to be a sea of answers, and I am being asked to locate one among them, while sifting water in my hands. Surely I am looking in the wrong part of the ocean, too, as how can I not be. In addition, I feel underequipped, in a knife-to-a-gunfight sense. How deeply I am persuaded of my inadequacies has to do with how many internal taunts I am willing to listen to and how long I will listen to them. Eventually I have to collect myself and reply, "You probably are going to get it wrong, but perhaps you will learn something." This doesn't quiet the objections, but it gives me somewhere to stand that is separate from them. I sometimes imagine myself as two different people. One entertains every anxiety provoked by the task, and the other, having handed off his distractions, tries to solve the problem with a less troubled mind. It gives me a little breathing room.

Occasionally at night there are formulas and equations in my dreams. These versions are likely nonsense, the thoughts of a man in solitary writing on a wall, at least I don't seem to have advanced any when I wake. It is simply as if I couldn't shake the concerns of the day. My mind appeared determined to keep after them, even if pointlessly.

Grasping something after long labor is sometimes a pleasure. Other times I feel indignant at having to spend so much time at a pursuit that might not be difficult for someone else and maybe even shouldn't have been so difficult for me. I feel now and then as if I am living in a different world from the people around me. Mine is rich but also fierce and exclusive and remote. They seem to have the run of creation.

SO FAR AS I can tell, the point of an education is to be introduced to books and art and matters of science and history and thought sufficient to engage one for the rest of one's life or to suggest a means of doing so. An education is a template for knowing where to look for the means to continue one's learning. Among the men and women whose classrooms I sat in were some inspiring figures and also some duds. Nearly all of them, however, concealed the notion, if they knew it, that I was engaged in a starter course of cultural appreciation as a means of preparing for a life of thought and response, and by thought I mean the capacity to reason one's way out of difficulties. One needs to know where to turn, to have direction. Children are weeded out from bright futures as often as creatures in the wild come to unhappy ends. No agency protects the innocent, Maxwell used to say. I understood late when I was young that within the approaching rush of the world, there was a path. I wish that I had understood it sooner. I might have learned more and felt confused less often. I might have seen opportunity instead of felt overwhelmed. This concern feeds an anxiety I have that I inhibited my interest, self-protectively, and that if I had challenged myself, instead of giving up, I might have more to show for it. There is only so much, though, that one can reasonably expect of one's childhood self.

SOMETHING ABOUT THE orderliness of calculus is what I imagine might have appealed to Maxwell, its compactness, its concision, its descriptive powers, its sweep, and its elegance. Amie thinks that he might have said that he loved its harmony. "I also think he found the rigorous aspect of it pleasing. How it presented a complete story that fit together so neatly," she said. Calculus resides near a border with a practical incoherence, the way language resides near a border with one's thoughts before they are entirely formed. Calculus is a means of personifying an abstraction. *The Happiness of Getting It Down Right* is the title of Maxwell's correspondence with the writer Frank O'Connor. It comes from a phrase Maxwell used to describe the pleasure of finding the right words to impose on a line of one's thoughts or feelings. Calculus allows a mathematician to clarify a movement or a quantity with an ideal specificity.

Algebra and geometry felt to me like attempts to characterize the world we can see and put our hands on. The prosaic observations of algebra leading to a refined view of circumstances you could almost but not quite figure out on your own. The simple, hard-looked-at, and by now scuffed-up shapes of Euclidean geometry, triangles and circles and squares, placed on pedestals and closely examined, the primary sculptural forms embedded in ordinary structures.

Calculus seems to enclose the world in its logic. It makes, as Berlinski writes, a circle. Algebra and geometry describe things that you see plainly or at one remove, so to speak. How far you are from something, how tall something else is, the flight of a bird making a pattern like a line on a graph. Calculus, Berlinski says, describes "the world's network of mathematical nerves."

Fall Again

1.

When I began, I didn't know how I would decide when I was done. I thought I might take the year-end, statewide exams for high school students, but the first question I read on the algebra exam asked me to solve a problem by using a formula that I had never heard of, and I decided, No good. I wasn't studying math as a competitive task, and I didn't want a grade as a measure of whether my work had been worthwhile.

Amie could tell where I was from the questions I asked. She would occasionally say, "You can move on. I think you've done enough." When I seemed stuck on the idea of a too-exquisite grasp of a concept or a procedure, she would say, sometimes with a mild exasperation, "Can't you move on now?"

2.

I was cut from the herd once more by a process called implicit differentiation. It wasn't clear to me if I didn't understand it, or whether I had failed to understand the steps that lead to it—that is, if I had fallen behind earlier and had, in effect, faked my way to it.

When I wrote Amie, she answered, "I'm sorry about ID. It's a multistep procedure. You have to keep track of what you're doing. For me it's very visual. I have to say, 'What's going on here? What does this mean?' There has to be a story that explains it." Her remarks reminded me of what it is to have an intelligence that is welcoming to such challenging material. I see only steps and abstruse rules, and she sees a story.

"I really think you should move on," she added. "You really are not going to finish in the foreseeable future at this pace."

By the age of sixty time no longer seems unlimited. At least that is when I first felt, obscurely, or now and then, that I am running out of time. A few years ago, the writer Roger Angell, who lately turned one hundred, told me, "When I was in my sixties, I thought about death all the time. Now I never think of it." I don't think about my end so much as I think that it is unreasonable to assume that the party will last forever. I know three landscapes intimately, two from my childhood and New York City, where I have lived for forty years. I feel close to no longer having sufficient time to learn another one. "I can only buy old guitars now," Ry Cooder, who is roughly my age, told me. "I haven't got the time anymore to break new ones in."

A STUDENT'S RELATION to the world is indirect, and only more so when the student is older. The things that occupy one's thinking do not bear on ordinary life. If they

apply, it is abstractly, in that they might make possible a
different future.

I review formulas and equations like an actor review-
ing lines. They occupy the forefront of my thinking. It is
strange to be older and about as skilled at something as a
child is likely to be.

3.

Mathematicians do not make irrational assertions about
mathematics, the way the rest of us do about things that
we believe are true. We are broadcasters and proclaimers
of self-sustaining remarks. It appears we never run out of
them. We deliver them even when we think we aren't.
The mathematics that is acknowledged to be true is aloof
from human bias and shortcomings. In mathematics no
one has to believe anything. There is no question of faith.
There are no math opinions about whether something is
true or not true.

Human beings deal mostly with incomplete or un-
proven truths. It would be very difficult, perhaps impossible, to construct an axiomatic catalog of truths that was
anything like comprehensive, because ideas of truth don't
agree. The standard of truth would seem to be absolute,
but in many cases it's only partial; what appears true to
me is not necessarily true to you. Struggling to understand a math problem is different from struggling to understand an argument. In math, when a voice within me
says, Why can't you understand this? I can't answer, Because I don't agree with it.

As I got older, I got better at reading, having read for many more years, of course, but also from maturing, of having developed the capacity to respond more deeply and in more complex ways, and of having worked to make more of myself than what I seemed able to be or fated to be. To grow beyond the restrictions of my raising. I had thought that some part of this greater capacity would lend itself to learning math. Learning math isn't necessarily congenial to outside influences, though, another way in which math is cold.

Thinking is a strange pursuit. You are making an approach on an idea. It doesn't want to stay where you can see it clearly or reveal all of itself to you at once. It doesn't hold a shape, it dissolves, it shows you different sides, it slips away. As hard as I try, before long I realize I am thinking of something else, and I have to start over and reconsider and reevaluate, only sometimes advancing. With calculus I am not infrequently confronted with ideas that I can't efficiently grasp. They seem just beyond my reach. There was a period near the end of my studying when Amie wasn't always taking my calls. I imagined her looking at her phone and thinking, Not now. The patience on the part of learned people for half-bright beginners is not inexhaustible.

Day in and out I deal with things that no one I know deals with. It limits my conversation. I can't in the evening say to a dinner companion, "We have a $g(x)$, what is the function that defines $f(g(x))$?" although it is a perfectly simple question.

4.

Ever tried. Ever failed. No matter. Try again. Fail again.
Fail better.

—Samuel Beckett, *Worstward Ho!*

Failure has become not a way station, as graduation
speakers sometimes describe it, or an obstacle to confront
and subdue, but a shadow. It never falls on me entirely,
but it never leaves me, either. Perhaps unconsciously and
self-destructively, I have chosen a task designed to show
that I am less than who I had believed myself to be. Perhaps I flattered myself about my abilities and have taken
a serious task lightly and now, having committed myself
to putting aside my regular life for more than a year, I
am discovering that I might fail for real. I might have
enacted something I believe, which is that most difficulties in life are the result of bad judgment. During these
periods, I see nothing romantic about failure. It feels
like a condition that has separated me from the people
around me; I look at others the way a sick person looks
at the well.

By Kathryn Schulz in her book *Being Wrong*, I was
made aware of Anne Carson's poem "Essay on What I
Think About Most," which begins,

> Error.
> And its emotions.
> On the brink of error is a condition of fear
> In the midst of error is a state of folly and defeat.
> Realizing you've made an error brings shame and
> remorse
> Or does it?

Failure as a companion makes me feel a little crazy, deluded in my purposes, like a monk who sees lower angels from the ones that everyone else in the monastery sees. For solace I looked for figures in myths and folk tales who fail repeatedly, but I couldn't find any. Sisyphus fails of course, but he is fulfilling a punishment, not attempting a task. Parsifal fails to ask the right question, but only once. Repeated failure doesn't appear to be a mythological subject.

Something I said to a friend, a philosopher, led him to write me:

> There are indeed some questions which are unanswerable, and some tasks which are unfulfillable, but many questions and tasks humans propose to themselves are answerable and fulfillable after some period of failure. Nevertheless, it is striking that many humans regard any failure as shameful. The mechanization of much of our lived world has led us not only to have unrealistic expectations of success in all endeavors but also to be both impatient with and frightened by failure. But regarding failure as shameful long precedes the industrial revolution; one of the reasons Socrates struck so many people as odd and even frightening was his lack of this attitude toward failure.

This made me feel encouraged to go on, in my own halting way.

IN FAILING, THOUGH, I had company. After I was nearly done with calculus and was downcast about my performance I heard Amie's husband, Benson, say one night at dinner, "I'm failing all the time. One time I was describing to someone how hard math is, because you're failing ninety percent of the time, and a mathematician happened to hear me and he said, 'Benson, you succeed ten percent of the time, you're amazing.' I said, 'No, I was exaggerating.'"

He shook his head. "Doing math is like I walk you to a wall, and it's a hundred feet high, and it's a sheer, straight wall, and it looks provably impossible to climb, but you have to climb it," he went on. "Everyone says it's not possible, and I say, 'Not with your current abilities. Your learning rate has to keep doubling. Levitate is the only strategy. You have to get to the point where you fly, and that's how you get over the wall.'"

Imagination is the biggest nation in the world, and in deep mathematics human beings can fly.

5.

At times, it is like walking into a room where the parts of a machine are on the floor, some of which you know how to put together, but you don't know how to make the whole machine, even though you think you understand what each part does, and meanwhile you're not sure what the capabilities of the completed machine are, either. Some you grasp, and some are to be revealed by its operation. The machine was put together for you the day before, or maybe only moments before, as a demonstration, but there are too many procedures to be consecutive in your mind, or some of them were skipped. They are discrete and until you find an order among them you will not know how to manage the difficulty.

In its higher ranges, mathematics involves a combination of literalness and abstraction. A formula might be abstract in its reasoning and claims, but it is intended to be used literally. "A good mathematician balances between rigorous logic and intuition," Deane told me. By the time a mathematician reaches the positions that he and Amie and Benson occupy, the path from one intuitive judgment to another is not always clear. You are now and then in the dark parts of the map, where the territory might be infinite. Leaps of imagination are required. You can leap to somewhere you don't recognize, though, while also aware that every step you took was correct. It is furthermore possible that others might have taken the same steps and arrived somewhere different from where you did. It is a little like giving directions to someone who tells you later that the directions were wrong, but you had told them the way to a place that you'd been to many times.

6.

I had imagined mathematics as a landscape and my studies as a series of tours and encounters, but travel books tend to be cheerful, and when I am despondent about math, I reflect, What the hell am I thinking? This has been so difficult, and I have grown so isolated that I no longer am confident that I understand the parameters of a social exchange—how much to reveal, how deeply to go into a subject, when to withhold a remark, the whole calculus of conversation. I've become a kind of mathematical shut-in. I don't go anywhere except in my mind. My favorite travel books—*In Patagonia, Great Plains, The Sudden View, Life on the Mississippi*—have weather and anecdotes and changes of scene, not to mention characters, the whole assembly moving the effort along as if in a procession. I keep repeating myself—math is hard, I thought I'd be better at it; math isn't flawed, I am—but it is because in an abstract realm where one has only oneself for company the scenery doesn't change appreciably very often, never mind the difficulty of having only a barely workable sense of the language and the landscape's being obscured from view in many places and in others being harsh and unwelcoming. On such terrain one can't help but be thrown back on oneself. For long periods, I have occupied the same ground day after day. I feel less the winsome and ingratiating traveler and more like the people I have read about who by a shipwreck or some other disaster were stranded in the Arctic. Each day repeated the day before, and the grueling task conducted amid a spiky landscape of ice that went all the way to the horizon in all directions. Of course I mean only metaphorically and hyperbolically. In those cases, though, each day the landscape looked unaltered, but it had been, subtly, because the ice had moved, because it was sitting on the sea.

Only later is it borne in on me that I am as if sitting on the sea, with the terrain shifting beneath me. The unconscious changes that are subject to different stimuli and a different timeline are not always immediately apparent.

I feel like I've been through a romance that has hit the rocks, though. Calculus is walking away, and I refuse to give up on it. I know that breakthroughs arrive after periods of confusion, and that I should welcome the

uncertainty; surely there is enough rhetoric in the culture
supporting the point, it's practically a T-shirt sentiment—
"Embrace the Confusion"—but I'm not always able to sit
easily with the anxiety. I remind myself that the ability to
bear anxiety is a trait of emotional well-being. I remind
myself of Keats's concept of negative capability—the artist's capacity, that is, for accepting "uncertainties, mysteries, doubts, without any irritable reaching after fact and
reason," which he wrote in a letter to his brothers in 1817.
The energy that ought to be available to me for learning,
though, is still often diverted into the service of an indignant resistance. I feel I ought to be able to manage it and
am annoyed that I can't. In such periods I think, What
am I, helpless? Then I have a response and a response to
the response and a response to that response until after a
while I have forgot where I was in my textbook, and I'm
just a disputatious cacophony of voices.

One morning, though, briefly, I have a respite of a kind,
maybe a fleeting insight. The frustration of doing calculations is replaced by my intuitively apprehending the
position the calculations occupy in a larger design, their
ability to make hidden things visible. Even though I often
can't perform them correctly, I see what they are meant to
do. I can see the outline of the plan, even if only shakily.

It got me so excited that I had to go outside and walk
around until I calmed down.

7.

When I came back in u substitution returned and my poetic
intuitions were demised. Once again calculus enumer-

ated all of my faults and the reasons we were ill-suited
and insisted that we see other subjects.

"U substitution is a tool," Amie said. "It's important,
it's used a lot, but it isn't fundamental to the discipline of
calculus."

She sensed I wasn't cheered. "Listen," she said, "you're
not going to learn calculus at a deep theoretical level. It
isn't possible in a year. You need to be familiar with its
workings, but that's all."

Then, "You should take a break. You've done really,
really, really a lot."

I wasn't sure if she thought that I had or was only trying to make me feel better, but I said, "If I do, here's what
will happen—"

"You'll forget the math."

"Yes."

"You won't forget."

"I suppose it's there, isn't it," I said, "but it doesn't really feel like it."

It's a strange thing I have done, locking myself away. I
didn't think it was strange when I began, because I didn't
foresee how long it would take or how relentless it would
be, or, I guess, how obsessively I would pursue it. I have
lain awake at night thinking about a lot of things, but I
never expected that among them would be mathematics.
No one I know has embarked on an undertaking that is
entirely interior and reclusive, so there is no one I can ask
about how it's supposed to go. It was meant to be a lark,
and it has become a reckoning. And lest I feel prideful, I
shouldn't forget that I'm secluded with disciplines that are
handled by adolescents.

"You have my permission to stop," Amie said.

I considered it, in order to return to a life where I am

something like competent. Where my mistakes aren't a referendum on my abilities. Where they are matters of inattention and haste, missed opportunities, and sometimes poor judgment, which is all that my mistakes in calculus are, but the imperious requirement of a correct answer is a remorseless imperative. In ordinary life, by means of circumstance or luck, I not infrequently escape the consequences. There isn't a daily record of my shortcomings. I think of a poem by William Meredith that I read when I was young, "Hazard Faces a Sunday in the Decline," in which Hazard is a painter in a difficult period. "The cat is taking notes against / his own household," Meredith writes. For me, this reckoning figure resonates.

Wherever I leave off will mean that there is far more that I don't understand than there is that I do understand. I had thought that there would be a clean break, that having studied junior varsity calculus I would have learned everything about mathematics that someone should be expected to know and could collect my gold star.

8.

A parting look at infinite things: to certain spiritual-thinking people in Russia whose beliefs were not orthodox, Cantor introduced a justification for their practices. By proposing novel infinities and naming previously incomprehensible ones, Cantor had brought these collections into being. In *Naming Infinity*, Loren Graham and Jean-Michel Kantor discuss Name Worshippers, who believed that repeating the name of God in the Jesus prayer, "Lord Jesus Christ, Son of God, have mercy on me, a sinner," brought them into a divine presence. In the same way, Michael Harris writes that Russian mathematicians believed that "mathematical objects were brought into being in the course of giving them names."

Naming makes an ineffable thing real. "Mathematicians sometimes bring into being objects that no one has ever thought of before," Loren Graham told me. "How do you know it exists? That it isn't something that you *think* exists, but you can't convince other people, like Cantor with set theory? The first step is giving it a name. In the Bible God says, 'Let there be light,' and, having named it, there is light, as if the naming had been the creating. How do I know there's a God? I can name him." Until Cantor named specific infinities, no one had thought that there might be more than one.

When Cantor was seventy-two, he went back into the hospital. Several times he wrote to his wife, asking to be allowed to come home. Instead, he died where he was, from a heart attack, in 1918.

9.

A quasi-mathematical means of considering God's existence is Pascal's wager, which is sometimes said to have introduced probability theory. A person making the wager has to choose a position among four that are offered. Either God exists, or He or She doesn't (or They don't, either), and one must decide whether or not to believe. If God exists, and you believe, you are saved. If God exists, and you don't believe, you lose. If God doesn't exist, and you believe, you denied yourself pleasures. If God doesn't exist, and you didn't believe, you didn't waste your time. Pascal concludes that the risks of not believing are greater than those of believing, so one might as well believe.

Arguments for the existence of God are called ontological arguments, concerned, that is, with the nature of being. The first ontological argument I know of was made in 1078 by St. Anselm, who describes God as "something than which nothing greater can be conceived." ("The true infinite or Absolute, which is in God, permits no determination," Cantor wrote in 1883.) Even a fool would agree that something that is greater than anything else that can be thought of must exist in the mind, since the fool understood the concept, Anselm writes. If it exists in the mind, it exists outside the mind, since such a circumstance would qualify as a greater state of existence than one embodied merely in thoughts. If it exists only in thoughts, then it is both something than which nothing greater can be conceived and something than which something greater can be conceived, a contradiction. Therefore, something greater than which nothing can be conceived exists both in the mind and in the world.

Gödel had an ontological proof, which he kept secret out of a concern that if people believed that he was a deist they might think less of him. In 1970, when he thought he was dying, he told a few people. He died in 1978, and his proof was published among a volume of his *Collected Papers*, in 1987.

In 2017, some German computer scientists said that they had run trials of Gödel's proof and that his assertions had held. Gödel's ontological proof is difficult and not understood entirely even by many philosophers, so I approach it cautiously. All I feel safe saying is that the computer program did not prove the existence of God; it proved that the form of logic that the proof involves, called modal logic, which is essentially an if/then logic, was consistent more or less with itself. To be persuaded of the proof's outcome, you would have to believe that all the propositions in the proof were true, and not everyone does.

Gödel's proof, which is an exemplification of St. Anselm's proof, using different methods, is usually stated in the following way:

Definition 1: x is godlike if and only if x has as essential properties those and only those properties which are positive

Definition 2: A is an essence of x if and only if for every property B, x has B necessarily if and only if A entails B

Definition 3: x necessarily exists if and only if every essence of x is necessarily exemplified

Axiom 1: If a property is positive, then its negation is not positive

Axiom 2: Any property entailed by—i.e., strictly implied by—a positive property is positive

Axiom 3: The property of being godlike is positive

Axiom 4: If a property is positive, then it is necessarily positive

Axiom 5: Necessary existence is positive

Axiom 6: For any property P, if P is positive, then being necessarily P is positive

Theorem 1: If a property is positive, then it is consistent, i.e., possibly exemplified

Corollary 1: The property of being godlike is consistent

Theorem 2: If something is godlike, then the property of being godlike is an essence of that thing
Theorem 3: Necessarily, the property of being godlike is exemplified

Extended to mathematics, this argument suggests an axiom that would answer all mathematical questions. Apprehending this axiom is likely outside of our abilities, the reasoning goes, but that does not mean that it doesn't exist. The only mind capable of creating such an axiom is God's. This agrees with Gödel's incompleteness theory in suggesting that for any system in which at least some arithmetic can be done there are statements that cannot be proved within the system. Two circumstances arise: the axioms will be inconsistent—that is, they might prove, for example, that $0 = 1$—or there will be certain true statements, such as $1 + 1 = 2$, that they won't be able to prove. It is also a Platonic notion, since it suggests that mathematics is the result of apprehending truths outside ourselves and that mathematics is not merely a complex structure built by human beings from arithmetic.

In suggesting an absolute explanation for mathematics Gödel arrived where Cantor had—there is an ultimate region in which mathematics resides, and it is located in the mind of God.

MATHEMATICIANS HAVE OBSERVED that unlike other arts, mathematics is the result of centuries of shared effort and in that way resembles a church as much as a creative discipline. I find these thoughts expressed by Robert Langlands in "Is There Beauty in Mathematical Theories?" a talk given at the University of Notre Dame in 2010.

Although I am not equipped to join the church of mathematics, I see that it could be a church. Per Langlands, like a church it represents the work of many people over centuries. There is a primary religious experience in the form of the transcendental engagement of doing mathematics; there are priests in the form of mathematicians who understand the mysteries, and sacred texts in the form of theorems, which instead of being inscrutable like proverbs are unambiguous, although requiring training and special knowledge to comprehend. To the initiates, though, their import is clear: God had a plan and He (or She or They) revealed it in the form of mathematics, and, so as to understand something of the plan's design and magnificence, the divinity gave us the intelligence to apprehend it, an idea that first appears in the Middle Ages.

You could also have a companion church that believes in the hegemony of beauty and the tenet that math is discovered not created and therefore enigmatic and suggestive of forces impossible to know, and you could have a schismatic branch that believes that mathematics is created by mathematicians and that beauty has nothing to do with it; nevertheless, mathematics is worth worshipping.

MY OWN EXPERIENCE with religion is limited. The only sacred text I am familiar with is the Bible, and I cannot separate it from my father. In the middle of his life, he became a Christian Scientist; I don't know why exactly. The questions one might ask of the dead pile up, and it is only one question I might ask him. He worked in an office on a high floor of a building in New York City, and from things he said later I pieced together the impression that

he had begun to feel deeply anxious when he had to pass
an open window. In addition, he had painful headaches.

Our family might have been better off if he had taken
a more worldly approach, if he had occupied a couch in
an analyst's office, say, but maybe not. Some people can't
take self-examination and collapse instead of getting better. Or shed their old lives for new ones, the way some
people survive car wrecks that kill everyone else. The
practice of Christian Science was a lean one, at least for a
child. You were made to feel, in a plainspoken way, that
if you fell short of the necessary faith, you would pay for
it with your well-being. You would get sick and become
crippled from polio and God would abandon you.

Part of my father's spiritual practice was to read from
the Bible and from *Science and Health* early in the morning, before leaving for the train that took him to the city.
One morning when I was four or five, I woke early and,
wandering downstairs, saw him on the couch at the far
end of the living room. A book was open on his lap, and
the light from the lamp beside him hid his face. A friend
of mine once said, "Your children never want to see you
scared." I wasn't aware that my father was scared, but I
had the feeling that he was doing something, not illicit,
perhaps, but solitary and obscurely desperate, and I was
unsettled. He kept the books in a drawer in a table beside
his bed, and I used to open the drawer and look at them,
especially the Bible, with its softbound leather cover and
marbled end pages, and wonder what he used it for, what
it did for him—that is, what secret did it contain?

The Bible is, among many other things of course, a
catalog of antique miracles and a plan for moral behavior.
Nearly every page contains a story about how to behave

or how not to. Only Dante, perhaps, spent as much time thinking of consequences as the writers of the Bible did. It's an enlarging book, of course, the way Shakespeare is enlarging, and Dante and Tolstoy. I can read the Bible partly as history or as a key to the states of mind of ancient people or as a guide to the complexities of belief. A good portion of it involves the attempt to find form for the magical. It can make you feel in the presence of holy events and holy people. It can reduce you to a state of wonder. A not inconsiderable number of people believe that the key to it is mathematics. I regard it as a work of art, a repository of human knowledge. A description of the terrible struggle to remain completely alive in the face of harsh circumstances.

TOO MUCH FAITH makes a person liable to being credulous, whereas too much intelligence might restrict faith. Faith that is humility, a belief that we don't know and might never know the larger design, the intention, the plan if there is one, seems sensible to me, but so does the belief that there is no plan.

It does not ask much of the imagination to regard mathematics, especially as it grows more abstruse, as a trail that followed to its end might bring one into the presence of God in the form of a super-axiom. From the figure described in the Old Testament it isn't difficult to picture a deity sufficiently pleased with His creation, even sufficiently vain, that He would like its design to be appreciated and so has left traces of its plan in the form of schematics. Or that there is a kind of endgame for humans in finally apprehending the design of the universe, a moment when paradise would arrive on earth. I'm not saying that I believe these things, only that a person can't say that they couldn't possibly be true.

In the eighteenth century, the belief that studying mathematics was a form of worship was exemplified by Maria Gaetana Agnesi, an Italian woman who wrote one of the first significant textbooks on calculus. According to a piece about Agnesi by Evelyn Lamb, published on the website of the Smithsonian, Agnesi was drawn to the thinking of the Jesuit philosopher Nicolas Malebranche, who wrote that "attention is the natural prayer of the soul." A Christian life involved strengthening one's intellect, and Agnesi considered the study of calculus to be a form of prayer.

The mathematician Leonhard Euler, in the eighteenth century, believed that imaginary numbers, such as the square root of −1, "are neither nothing, nor less than nothing, which necessarily constitutes them imaginary, or impossible." To Leibniz they were "a fine and wonderful refuge of the Holy Spirit, a sort of amphibian between being and not being." Leibniz invented the dyadic or binary system, in which all numbers are represented by 0 and 1 and combinations of them. Leibniz equated 0 with nothing, and 1 with God, who had created all things out of nothing. Leibniz was so excited about his system that he asked the president of the mathematical tribunal to China, a Jesuit named Grimaldi, to describe it to the emperor of China. This is reported in the paper "On the Representation of Large Numbers and Infinite Processes," by Arnold Emch, which was published in 1916. "Leibniz hoped that in this manner the Chinese Emperor might be won over to Christianity," Emch writes.

10.

A modern example of a distinguished mathematician who was also a God-seeker is Alexander Grothendieck, who died in 2014. I know about him from Deane, who revered him. Grothendieck was an algebraic geometer, meaning that he studied equations that describe physical spaces. Michael Harris describes Grothendieck's work as "part of a search for total purity." Another writer said of Grothendieck that following his work one has the "impression of rising step-by-step towards perfection. The face of Buddha is at the top, a human, not a symbolic· face, a true portrait and not a traditional representation." In 1988, Grothendieck fasted for forty-five days, intending to force God to show himself.

Grothendieck was born in Berlin in 1928 and died in a small town in France called Ariège, in the Pyrenees. According to Winfried Scharlau, a German mathematician who has written a biography of him, Grothendieck lived a "very unusual life on the fringes of human society." François Hollande, the president of France, described Grothendieck when he died as "an out-of-the-ordinary personality in the philosophy of life."

In the late 1950s and early '60s, Grothendieck was married and had three children, as well as a son from a woman he wasn't married to. He sometimes let people who were down on their luck stay at his house for weeks at a time. He taught in Brazil, and at the University of Kansas, and he lectured at Harvard. In 1966, Grothendieck received the Fields Medal, the most prestigious award in mathematics, but to protest the Soviet Union's having imprisoned two writers, he refused to go to Moscow to accept it. In 1970, when he was forty-two, he quit mathematics altogether.

He became a Buddhist, but in 1980 he took up a type of Christian mysticism, and for many nights in a row he "played chorales on the piano and sang," Scharlau writes. In 1991, Grothendieck secluded himself and worked on his mathematical meditations. According to Scharlau, these "cover biographical, religious, esoteric, and philosophical themes," but they haven't been translated. By the time Grothendieck died he had written thousands of pages. His obituary in *The New York Times*, written by Bruce Weber and Julie Rehmeyer, quotes the following remarks from his writings: "Among the thousand-and-one faces whereby form chooses to reveal itself to us, the one that fascinates me more than any other and continues to fascinate me, is the structure hidden in mathematical things."

11.

Plato's conception of mathematics was metaphysical more than theological. He believed that mathematical truths made the mind receptive to higher truths and an awareness of the Good, an entity different from the Greek gods in that it didn't look like a person or have supernatural attributes. The Good was the divine force that organized ideas. For later thinkers, understanding that terms of infinity applied in mathematics enabled a person to understand the expansive quality of truth and its place as a concept central to a higher form of existence. The sixteenth-century mathematician Giordano Bruno saw mathematics as a link between the terrestrial and celestial worlds.

One night at dinner, toward the end of my studying, Amie said that, so far as she could tell from the discussions we had had, my book might turn out to be interesting, but she hoped it wouldn't have too much about Platonism, which startled me sufficiently that I was unable to frame a reply. She is, and I guess I was surprised to learn, a fairly vehement anti-Platonist. The home ground of her discoveries is not a non-spatiotemporal realm, she said, a trifle emphatically for my comfort, but an abstract one created by mathematicians.

"Human beings have shone a light on numbers, and we've picked out a logical system," she said. "You can't bring God into this. It's unnecessary."

After that I shut up around her about Plato.

She also said, "I have to admit I was kind of alarmed when I realized how bad your arithmetic skills were."

"How did you know that?"

"From the things you would ask."

12.

Mathematics did not embrace me. It tolerated me. With practice, I developed a low-grade proficiency at finding derivatives and antiderivatives and factoring polynomials, but those are only methods. I was pleased with being able most of the time to describe how fast the train from Omaha was traveling as it passed through Kansas, but that simple competence didn't resemble the capacity to have thoughts in another language. There is always a gap between mere ability and the invocation of insight or imagination. Of course, I didn't really expect to get there.

I didn't even expect to get to where I might understand one of Amie's papers. Although I had viewed calculus as an end in itself, it isn't; it's simply an acquisition of practices, a station on the way to an unspecified front.

"Innocently to amuse the imagination in this dream of life is wisdom," Oliver Goldsmith writes in *The Vicar of Wakefield*, published in 1766. If one is fortunate, and sometimes I was with mathematics, more by reading about its ideas and its reach than by doing it, but if one is fortunate, one has something that engages one's attention, a diversion that makes one lose the feeling of time's passing, as in childhood. Time passes as it does in childhood, because one is adrift among the senses. There is only narrative—this happened, then this happened, and after that . . . Adults contribute the notion that there isn't enough time.

I'm not sure what I thought I would accomplish, although I hoped to achieve something more than a fair to middling facility at math. My ambition had been to expose myself to something challenging and to become different from who I was. I didn't know if things might change gradually or suddenly, or how they might change, or if they would change at all. Just because I wanted them to didn't mean that they would.

The things that were happening to me, though, were happening somewhere else than in the forefront of my thinking and differently from the way that I had imagined they would happen. What I had thought would be a process of enlarging was instead often a process of concentrating. As a practical matter, I had learned to solve problems tactically, by viewing them as combinations of parts, which was broadly valuable. Meanwhile, though, certain habitual manners and patterns of undermining myself, certain hesitancies, even a kind of self-supporting narcissism, necessary to sustain myself against threats of inconsequence, appeared to have relaxed their hold. Having been shown day in and out not to be useful, they seemed to have withdrawn, quietly, without announcement, and without my having noticed their leaving. Apparently I had enacted Judith Rodin's belief that mastering a task later in life makes a person more confident. I hadn't mastered my task, but I had at least kept at it.

I began then to regard failure differently. I had thought of my failures as setbacks amounting to a nearly perpetual defeat, but perhaps they also had other meanings and consequences. Failure as an emotional event was disruptive and in the day-after-day sense was static and oppressive, an affront to my vanity, but failure as a psychic event appeared to be an incitement, a provocation of the what-are-you-going-to-do-about-it kind. In Hinduism there is the story of the churning of the ocean of milk in order to obtain amrita, the nectar of eternal life. The ocean delivers several valuable gifts, including amrita, but the first thing it delivers is a poison strong enough to demolish the world. Shiva, the god who destroys and creates the world, drinks the poison, which turns his neck blue. Shiva destroys things so that they can be re-created in a better form, illusions as well as material things. In challenging myself, perhaps I had been about demising illusions that I clung to. Or perhaps in failing I had simply been discovering what it means to allow things to fall apart.

13.

On the night at the dinner table where Amie shut me up about Platonism, Benson said that doing mathematics was for him a "quasi-religious experience." He has been working for several years on a problem called Hilbert's thirteenth problem, which has been open at least since Hilbert proposed it in 1900 but also has elements of problems first studied by the Babylonians in 3000 BCE. (It involves seventh-degree polynomials—that is, equations of the form $X^7 + aX^6 + bX^5 + cX^4 + dX^3 + eX^2 + fX + g = 0$.)

Everything else fell aside when he worked, he said, and he found himself feeling connected to people who were thinking of the same things hundreds of years ago. "It's not like you're talking to God, it's like you *are* a god," he said. As a boy he had been a tennis player with a regional ranking, but doing anything but math had come to seem trivial and silly, although he read at night because mathematics often got him too wound up to sleep.

I asked when he decided to be a mathematician, and he said, "There's a great mathematician named Dennis Sullivan, and one time we're with a bunch of people, at a conference in Helsinki, we're on a boat, and he asked us, 'When did you know you wanted to be a mathematician?' Some people said, 'When I was in grad school.' I said, 'I can name the day when I knew I wanted to become a mathematician.' So Dennis Sullivan looked me up and down, and he said, 'You were fourteen.' I said, 'How in the world could you possibly know this?' and he said that either he has a theory, or he heard of a theory, that the moment that one wants to become a mathematician, all emotional maturity ends. And he said, 'Don't feel bad, Benson, I was thirteen.'"

I wish that I were able to invent such an exchange, because it would suggest that while I hadn't been any good
at mathematics, I had nevertheless absorbed something
of its essence.

14.

The incremental way that calculus reckons change suggests the patterns by which our lives advance, either years
at a time or slight change upon slight change, much of it
either so slight or so grand as to elude ordinary notice. The
movement from one second to another. The revolutions of
night and day. The years that pass more quickly than we
can account for. The inherent motion of life, its rushing
river quality. Our lives progress, infinitesimally, but also
majestically, one breath, one heartbeat after another, the
increases accumulating while the range grows smaller,
with perfection as one limit and death as another. If our
existences are described by words, they are also described
by numbers in their pure states. It is a further application
of the unreasonable effectiveness of mathematics, one that
is a little chastening to acknowledge. This led me to wonder if what Maxwell loved about calculus was its majestic
precision, its beauty, its God's eye view.

I WROTE AMIE with a calculus question, and she said that
what I was trying to learn was considerably beyond intro-
ductory calculus and useful only if I were trying to visit
the discipline's higher ranges. "You have my permission
to stop," she wrote, so I did.

15.

I did not foresee that learning adolescent math would lead
me to notions of divinity. In my defense I will point out
that I did not blaze a trail; I followed footprints worn into
history. As a child I sometimes had the sense of an accompanying presence, of something immaterial behind
everything. It wasn't a thought so much as an intuition,
a sense of being in the company of, a proximity. This
manner of thinking is called Immanence, in which the divine is believed to be among us, as it were, sensed but not
seen. So many people have experienced this—after all,
it has a name—that it seems quaint to regard it as original or even unusual. I simply add myself to the company
of those who have felt it. There is a period of childhood
when the balance between conscious awareness and the
Unconscious is not weighted so much toward consciousness, when one receives sensations differently and is less
inclined to examine them. Everything is too immediate
and too novel for examination, and anyway one hasn't yet
developed any examining faculties.

Why have I found this subtext of God knowing so interesting? Partly because it was unexpected, the notion
that numbers are hidden in the world and the divine might
be hidden in numbers, but also because, like many people,
I want to believe in something more than the ordinary
terms of life. It's an ancient human longing, as anyone
knows, and there is solace in being a member of a benign
and well-wishing human community. You'll die alone is
an insult and a threat. I have found it inspiring to share the
company of large thinkers. The non-sneerers. The gropers
toward knowledge. The knowledge fluent. The failers and
perseverers, and the substantial and vulnerable women

and men in all cultures and places and of all races who laid themselves open to inspiration, mystery, and joy. They have raised my spirits. They have extended the boundaries of the lived world, and although I can't be one of them, I can cheer them on and exult in their accomplishments.

Even if mathematics didn't seem to want to have much to do with me, I see no reason to nurture a grievance. I can still celebrate it as an idea carried down through history like a sacred knowledge by so many different minds, durable and adaptable and only partly explored and some of it, perhaps even much of it, presently out of reach entirely. It wasn't my intention to become good at it, I wanted to submit myself to it and allow myself to respond to what it acquainted me with. Surely I wish I had done better at it, but it taught me not to be complacent in my thinking. An experience that only flattered my vanity would have taught me nothing at all.

Studying mathematics made me aware of a natural structure, elusively apparent and perhaps ultimately impenetrable. An implicit orderliness. An unfolding, moment by moment, on an apparently spectacular scale of something that no force can interrupt, something that is perhaps force itself. A trembling quality to life, both fearsome and fragile, a pattern that even to a novice like me is as clear as the grain in a piece of wood. I am aware that this way of regarding the world might be seen as unoriginal and romantic, but I find it consoling nevertheless, and mathematics, by my slight acquaintance, delivered me to it, which is more than I had asked that it do.

I am pleased for what learning of a pure type has done for me. I say pure meaning that I have no means of using algebra, geometry, or calculus; I'm not trying to get into college or preparing for a new career. Despite my resistance and my incapacity, mathematics broadened me. By the close company of a discipline that insisted that I think and reason, I was enlarged. Like the philosopher at the dinner table, I understand the value of inquiry now and am more inclined to listen and less inclined to resist or pronounce. I assume that a problem has dimensions that I haven't yet grasped or am even unaware of and that only a receptive examination can advance my understanding, meanwhile knowing—what I didn't before—that all thought, all knowledge, all opinions and beliefs are everlastingly subject to revision.

I am grateful to have learned this. I wish I knew who to express this gratitude to.

BIBLIOGRAPHY

BOOKS

Acheson, David. *The Calculus Story.* Oxford: Oxford University Press, 2017.

Aguirre, Anthony, Brendan Foster, and Zeeya Merali, eds. *Trick or Truth? The Mysterious Connection Between Physics and Mathematics.* Cham, Switzerland: Springer International Publishing, 2016.

Baddeley, A. D. *Working Memory.* New York: Oxford University Press, 1986.

Balaguer, Mark. *Platonism and Anti-Platonism in Mathematics.* New York: Oxford University Press, 1998.

Bellos, Alex. *The Grapes of Math: How Life Reflects Numbers and Numbers Reflect Life.* New York: Simon & Schuster, 2014.

Berlinski, David. *A Tour of the Calculus.* New York: Vintage Books, 1997.

Birrell, Anne. *Chinese Myths.* Austin: University of Texas Press, 2000.

Bransford, J. D. *Human Cognition: Learning, Understanding, and Remembering.* Belmont, CA: Wadsworth, 1979.

Bronowski, Jacob. *The Ascent of Man.* London: Book Club Associates, 1973.

Brown, James Robert. *Philosophy of Mathematics: A Contemporary Introduction to the World of Proofs and Pictures,* 2nd ed. New York: Routledge, Taylor & Francis Group, 2008.

Brown, Richard G., Ray C. Jurgenson, and John W. Jurgenson. *Geometry.* Evanston, IL: McDougal Little, 2011.

Caldwell, Chris, and G. L. Honaker, Jr. *Prime Curios! The Dictionary of Prime Number Trivia.* CreateSpace, 2009.

Carey, Benedict. *How We Learn: The Surprising Truth About When, Where, and Why It Happens.* New York: Random House, 2015.

Changeux, Jean-Pierre, and Alain Connes. *Conversations on Mind, Matter, and Mathematics.* Princeton, NJ: Princeton University Press, 1995.

Cheng, Eugenia. *Beyond Infinity: An Expedition to the Outer Limits of Mathematics*. New York: Basic Books, 2017.

Conway, John H., and Richard K. Guy. *The Book of Numbers*. New York: Copernicus, 1996.

Courant, Richard, and Herbert Robbins. *What Is Mathematics? An Elementary Approach to Ideas and Methods*. New York: Oxford University Press, 1996.

Darling, David, and Agnijo Banerjee. *Weird Math: A Teenage Genius and His Teacher Reveal the Strange Connections Between Math and Everyday Life*. New York: Basic Books, 2018.

Davis, Philip J., and Reuben Hersh. *The Mathematical Experience*. New York: Mariner Books, 1998.

Dedekind, Richard. *Essays on the Theory of Numbers*. Chicago: Open Court Publishing Company, 1901.

Dehaene, Stanislas. *The Number Sense: How the Mind Creates Mathematics*. New York: Oxford University Press, 1997.

Descartes, René. *Discourse on the Method of Rightly Conducting the Reason and Seeking the Truth in the Sciences*. New York: P. F. Collier & Son, 1909.

Ellenberger, Henri F. *The Discovery of the Unconscious: The History and Evolution of Dynamic Psychiatry*. New York: Basic Books, 1970.

Euclid. *Euclid's Elements*. Translated by Thomas L. Heath. Santa Fe, NM: Green Lion Press, 2017.

Franz, Marie-Louise von. *Alchemy: An Introduction to the Symbolism and the Psychology*. Toronto: Inner City Books, 1980.

Frenkel, Edward. *Love and Math: The Heart of Hidden Reality*. New York: Basic Books, 2013.

Geary, D. C. *Children's Mathematical Development: Research and Practical Applications*. Washington, DC: American Psychological Association, 1994.

Gelfand, I. M., and Alexander Shen. *Algebra*. Basel and Boston: Birkhäuser Verlag, 2004.

Gowers, Timothy. *Mathematics: A Very Short Introduction*. New York: Oxford University Press, 2002.

Graham, Loren, and Jean-Michel Kantor. *Naming Infinity: A True Story of Religious Mysticism and Mathematical Creativity*. Cambridge, MA: Harvard University Press, 2009.

Hacking, Ian. *Why Is There Philosophy of Mathematics at All?* Cambridge and New York: Cambridge University Press, 2014.

Hardy, G. H. *Ramanujan: Twelve Lectures on Subjects Suggested by His Life and Work*. Providence: AMS Chelsea Publishing, 1999.

Harris, Michael. *Mathematics Without Apologies: Portrait of a Problematic Vocation*. Princeton, NJ: Princeton University Press, 2017.

Heaton, Luke. *A Brief History of Mathematical Thought*. New York: Oxford University Press, 2017.

Hoffman, Paul. *The Man Who Loved Only Numbers: The Story of Paul Erdős and the Search for Mathematical Truth*. New York: Hyperion, 1998.

Hofstadter, Douglas J. *Gödel, Escher, Bach: An Eternal Golden Braid*. New York: Basic Books, 1999.

Iamblichus. *The Theology of Arithmetic: On the Mystical, Mathematical and Cosmological Symbolism of the First Ten Numbers*. Translated by Robin Waterfield. Grand Rapids, MI: Phanes Press, 1988.

Kasner, Edward, and James R. Newman. *Mathematics and the Imagination*. Mineola, NY: Dover Publications, 1940.

Katz, Victor J., and Karen Hunger Parshall. *Taming the Unknown: A History of Algebra from Antiquity to the Early Twentieth Century*. Princeton, NJ: Princeton University Press, 2020.

Kleppner, Daniel, and Norman Ramsey. *Quick Calculus: A Self-Teaching Guide*. New York: John Wiley & Sons, 1985.

Kline, Morris. *Mathematics: The Loss of Certainty*. New York: Fall River Press, 1980.

———. *Mathematics for the Nonmathematician*. New York: Dover Press, 1967.

Lakoff, George, and Rafael E. Nuñez. *Where Mathematics Comes From: How the Embodied Mind Brings Mathematics into Being*. New York: Basic Books, 2000.

Lang, Serge. *The Beauty of Doing Mathematics: Three Public Dialogues*. New York: Springer Verlag, 1985.

Larson, Ron, Laurie Boswell, Timothy Kanold, and Lee Stiff. *Algebra 1*. Orlando: Holt McDougal, 2012.

Livio, Mario. *Is God a Mathematician?* New York: Simon & Schuster, 2009.

Lockhart, Paul. *Arithmetic*. Cambridge, MA: Harvard University Press, 2017.

Manin, Yuri I. *Mathematics as Metaphor: Selected Essays of Yuri I. Manin*. Providence: American Mathematical Society, 2007.

Mazur, Barry. *Imagining Numbers: (particularly the square root of minus fifteen)*. New York: Farrar, Straus and Giroux, 2003.

Nickerson, Raymond S. *Mathematical Reasoning: Patterns, Problems, Conjectures, and Proofs*. New York: Psychology Press, 2010.

Oakley, Barbara A. *A Mind for Numbers: How to Excel at Math and Science (Even if You Flunked Algebra)*. New York: Jeremy P. Tarcher/Penguin, 2014.

Paulos, John Allen. *Innumeracy: Mathematical Illiteracy and Its Consequences*. New York: Hill and Wang, 1988.

——. *Irreligion: A Mathematician Explains Why the Arguments for God Just Don't Add Up.* New York: Hill and Wang, 2008.

Penrose, Roger. *The Road to Reality: A Complete Guide to the Laws of the Universe.* New York: Vintage Books, 2007.

Polo, Marco. *The Travels of Marco Polo.* New York: Penguin Putnam, 1958.

Pólya, George. *How to Solve It: A New Aspect of Mathematical Method.* Princeton, NJ: Princeton University Press, 1945.

Ruelle, David. *The Mathematician's Brain.* Princeton, NJ: Princeton University Press, 2007.

Russell, Bertrand. *Mysticism and Logic and Other Essays.* London: George Allen & Unwin Ltd., 2010.

Sautoy, Marcus du. *The Music of the Primes: Searching to Solve the Greatest Mystery in Mathematics.* New York: Harper Perennial, 2003.

Shimura, Goro. *The Map of My Life.* New York: Springer Science-Business Media, 2008.

Spinoza, Benedict de. *Ethics.* New York: Penguin Putnam, 1995.

Steen, Lynn Arthur, ed. *Mathematics Today: Twelve Informal Essays.* New York: Springer Verlag, 1978.

Sterling, Mary Jane. *Algebra for Dummies.* Hoboken, NJ: John Wiley & Sons, 2010.

Stewart, Ian. *The Beauty of Numbers in Nature: Mathematical Patterns and Principles from the Natural World.* Cambridge, MA: MIT Press, 2017.

Strogatz, Steven. *The Joy of X: A Guided Tour of Math, from One to Infinity.* Boston: Houghton Mifflin Harcourt, 2012.

Taylor, Thomas. *Theoretic Arithmetic of the Pythagoreans.* New York: S. Weiser, 1972 (1st ed. 1816).

Thompson, Silvanus. *Calculus Made Easy.* London: Macmillan, 1914.

Watkins, Matthew. *Secrets of Creation.* Vol. 1: *The Mystery of the Prime Numbers.* Charlotte, NC: John Hunt, 2015.

Weil, André. *The Apprenticeship of a Mathematician.* Translated by Jennifer Gage. Basel and Boston: Birkhäuser Verlag, 1992.

Weinberg, Steven. *Dreams of a Final Theory.* New York: Vintage Books, 1993.

Wybrow, Cameron, ed. *Creation, Nature, and Political Order in the Philosophy of Michael Foster (1903–1959): The Classic Mind Articles and Others, with Modern Critical Essays.* Lewiston, NY: Edwin Mellen Press, 1992.

ARTICLES

Aloğlu, Mustafa, and Paul Zelhart. "Psychometric Properties of the Revised Mathematics Anxiety Rating Scale." *Psychological Record* 57, no. 4 (2007): 593–611.

Amalric, Marie, and Stanislas Dehaene. "Origins of the Brain Networks for Advanced Mathematics in Expert Mathematicians." *Proceedings of the National Academy of Sciences* 113, no. 18 (May 2016): 4909–17.

Andrews, G. E. "An Introduction to Ramanujan's 'Lost' Notebook." *American Mathematical Monthly* 86 (1979): 89–108.

Ashcraft, M. H., and K. S. Ridley. "Math Anxiety and Its Cognitive Consequences: A Tutorial Review." In *Handbook of Mathematical Cognition*. Edited by Jamie I. D. Campbell. New York: Psychology Press, 2005, 315–27.

Attig, M. S., and L. Hasher. "The Processing of Occurrence Information by Adults." *Journal of Gerontology* 35 (1980): 66–69.

Baddeley, A. D., and G. J. Hitch. "Working Memory." In *The Psychology of Learning and Motivation*, vol. 8. Edited by G. H. Bower. New York: Academic Press, 1974, 47–90.

Bartelet, D., et al. "What Basic Number Processing Measures in Kindergarten Explain Unique Variability in First-Grade Arithmetic Proficiency?," *Journal of Experimental Child Psychology* 117 (2014): 12–28.

Bean, J., and M. Metzner. "A Conceptual Model of Nontraditional Undergraduate Student Attrition." *Review of Educational Research* 55 (1985): 485–540.

Beilock, S. L., et al. "Female Teachers' Math Anxiety Affects Girls' Math Achievement." *Proceedings of the National Academy of Sciences* 107 (2010): 1860–63.

Beilock, S. L., et al. "More on the Fragility of Performance: Choking Under Pressure in Mathematical Problem Solving." *Journal of Experimental Psychology: General* 133 (2004): 584–600.

Bishop, S. J. "Neurocognitive Mechanisms of Anxiety: An Integrative Account." *Trends in Cognitive Sciences* 11, no. 7 (2007): 307–16.

Bradford, J. D., and M. K. Johnson. "Contextual Prerequisites for Understanding: Some Investigations of Comprehension and Recall." *Journal of Verbal Learning and Verbal Behavior* 11 (1972): 717–26.

Burke, D. M., and L. L. Light. "Memory and Aging: The Role of Retrieval Processes." *Psychological Bulletin* 90, no. 3 (1981): 513–46.

Cajori, Florian. "Absurdities Due to Division by Zero: An Historical Note." *Mathematics Teacher* 22, no. 6 (1929): 366–68.

Cassou-Noguès, Pierre. "On Gödel's 'Platonism.'" *Philosophia Scientiae* 2 (September 2011): 137–72.

Chavez, C. "Involvement, Development, and Retention: Theoretical Foundations and Potential Extensions for Adult Community College Students." *Community College Review* 34 (2006): 139–52.

Cho, S., et al. "How Does a Child Solve 7 + 8? Decoding Brain Activity Patterns Associated with Counting and Retrieval Strategies." *Developmental Science* 14 (2011): 989–1001.

Coffey, M. "Rising to the Numeracy Challenge." *Adults Learning* 22 (2011): 30–31.

Connelly, S. L., L. Hasher, and R. T. Zacks. "Age and Reading: The Impact of Distraction." *Psychology and Aging* 6 (1991): 533–41.

DeCaro, M. S., et al. "Diagnosing and Alleviating the Impact of Performance Pressure on Mathematical Problem Solving." *Quarterly Journal of Experimental Psychology* 63 (February 2010): 1619–30.

Dehaene, S., et al. "Sources of Mathematical Thinking: Behavioral and Brain-Imaging Evidence." *Science* 284 (1999): 970–74.

De Smedt, B., I. D. Holloway, and D. Ansari. "Effects of Problem Size and Arithmetic Operation on Brain Activation During Calculation in Children with Varying Levels of Arithmetical Fluency." *Neuroimage* 57 (2011): 771–81.

Dumontheil, I., and T. Klingberg. "Brain Activity During a Visuospatial Working Memory Task Predicts Arithmetical Performance 2 Years Later." *Cerebral Cortex* 22 (2012): 1078–85.

Faust, M. W., M. H. Ashcraft, and D. E. Fleck. "Mathematics Anxiety Effects in Simple and Complex Addition." *Mathematical Cognition* 2 (1996): 25–62.

Fennema, E. "The Study of Affect and Mathematics: A Proposed Generic Model for Research." In *Affect and Mathematical Problem Solving: A New Perspective*. Edited by D. B. McLeod and V. M. Adams. New York: Springer Verlag, 1989, 205–19.

Freeston, M. H., et al. "Cognitive Intrusions in a Non-Clinical Population. I. Response Style, Subjective Experience, and Appraisal." *Behaviour Research and Therapy* 29 (1991): 585–97.

Galla, B. M., and J. J. Wood. "Emotional Self-Efficacy Moderates Anxiety-Related Impairments in Math Performance in Elementary School–Aged Youth." *Personality and Individual Differences* 52 (2012): 118–22.

———. "Cognitive Predictors of Achievement Growth in Mathematics: A 5-Year Longitudinal Study." *Developmental Psychology* 47 (2012): 1539–52.

———. "From Infancy to Adulthood: The Development of Numerical Abilities." *European Child and Adolescent Psychiatry* 9 (2000): II/11–II/16.

———. "Mathematical Disabilities: Cognitive, Neuropsychological, and Genetic Components." *Psychological Bulletin* 114 (1993): 345–62.

Ginsburg, L. "Effective Strategies for Teaching Math to Adults." In B. H. Wasik, ed., *Handbook of Family Literacy*, 2nd ed. New York: Routledge, 2012, 195–209.

Hembree, R. "The Nature, Effects, and Relief of Mathematics Anxiety." *Journal for Research in Mathematics Education* 21 (1990): 33–46.

Hopko, D. R., et al. "Mathematics Anxiety and Working Memory: Support for the Existence of a Deficient Inhibition Mechanism." *Journal of Anxiety Disorders* 12 (1998): 343–55.

Hopko, D. R., et al. "The Emotional Stroop Paradigm: Performance as a Function of Stimulus Properties and Self-Reported Mathematics Anxiety." *Cognitive Therapy and Research* 26 (2002): 157–66.

Huffman, Carl. "Pythagoras." In the Stanford Encyclopedia of Philosophy (Winter 2018). Edited by Edward N. Zalta. https://plato.stanford.edu/archives/win2018/entries/pythagoras/>.

Hunsley, J. "Cognitive Processes in Mathematics Anxiety and Test Anxiety: The Role of Appraisals, Internal Dialogue, and Attributions." *Journal of Educational Psychology* 79 (1987): 388–92.

Hunt, T. E., D. Clark-Carter, and D. Sheffield. "The Development and Part-Validation of a U.K. Scale for Mathematics Anxiety." *Journal of Psychoeducational Assessment* 29 (2011): 455–66.

———. "Math Anxiety, Intrusive Thoughts and Performance: Exploring the Relationship Between Mathematics Anxiety and Performance: The Role of Intrusive Thoughts." *Journal of Education, Psychology and Social Sciences* 2, no. 2 (2014): 69–75.

Jameson, Molly M., and Brooke R. Fusco. "Math Anxiety, Math Self-Concept, and Math Self-Efficacy in Adult Learners Compared to Traditional Undergraduate Students." *Adult Education Quarterly* 64, no. 4 (2014): 306–22.

Jones, W. J., T. L. Childers, and Y. Jiang. "The Shopping Brain: Math Anxiety Modulates Brain Responses to Buying Decisions." *Biological Psychology* 89 (2012): 201–13.

Kasworm, C. E. "Emotional Challenges of Adult Learners in Higher Education." *New Directions for Adult and Continuing Education* 120 (2008): 27–34.

Kellogg, J. S., D. R. Hopko, and M. H. Ashcraft. "The Effects of Time Pressure on Arithmetic Performance." *Journal of Anxiety Disorders* 13 (1999): 591–600.

Linsky, B., and E. Zalta. "Naturalized Platonism Versus Platonized Naturalism." *Journal of Philosophy* 92 (1995): 525–55.

Lyons, I. M., and S. L. Beilock. "Mathematics Anxiety: Separating the Math from the Anxiety." *Cerebral Cortex* 22, no. 9 (2012): 2102–10.

————. "When Math Hurts: Math Anxiety Predicts Pain Network Activation in Anticipation of Doing Math." *PLoS ONE* 7(10): e48076.doi:10.1371/journal.pone.0048076.

Ma, X. "A Meta-Analysis of the Relationship Between Anxiety Toward Mathematics and Achievement in Mathematics." *Journal for Research in Mathematics Education* 30 (1999): 520–40.

Mazur, Barry. "Mathematical Platonism and Its Opposites." *European Mathematical Society Newsletter* 68 (2008): 19–21.

McMullan, M., R. Jones, and S. Lea. "Math Anxiety, Self-Efficacy, and Ability in British Undergraduate Nursing Students." *Research in Nursing & Health* 35 (2012): 178–86.

Miller, H., and J. Bischel. "Anxiety, Working Memory, Gender, and Math Performance." *Personality and Individual Differences* 37 (2004): 591–606.

Mozolic, Jennifer L., Satoru Hayasaka, and Paul J. Laurienti. "A Cognitive Training Intervention Increases Resting Cerebral Blood Flow in Healthy Older Adults." *Frontiers in Human Neuroscience* 4, no. 16 (2010).

Munoz, E., et al. "Intrusive Thoughts Mediate the Association Between Neuroticism and Cognitive Function." *Personality and Individual Differences* 55 (2013): 898–903.

Nixon, R., et al. "Metacognition, Working Memory, and Thought Suppression in Acute Stress Disorder." *Australian Journal of Psychology* 60 (2008): 168–74.

Patrick, H. "The Classroom Environment and Students' Reports of Avoidance Strategies in Mathematics: A Multimethod Study." *Journal of Educational Psychology* 94 (2002): 88–106.

Poldrack, R. A. "Can Cognitive Processes Be Inferred from Neuroimaging Data?," *Trends in Cognitive Sciences* 10 (2006): 59–63.

Price, G. R., M. M. Mazzocco, and D. Ansari. "Why Mental Arithmetic Counts: Brain Activation During Single Digit Arithmetic Predicts High School Math Scores." *Journal of Neuroscience* 33 (2013): 156–63.

Richardson, F. C., and R. M. Suinn. "The Mathematics Anxiety Rating Scale." *Journal of Counseling Psychology* 19 (1972): 551–54.

Rosenberg-Lee, M., M. Barth, and V. Menon. "What Difference Does a Year of Schooling Make? Maturation of Brain Response and Connectivity Between 2nd and 3rd Grades During Arithmetic Problem Solving." *Neuroimage* 57 (2011): 796–808.

Rowett, Catherine. "Philosophy's Numerical Turn: Why the Pythagoreans' Interest in Numbers Is Truly Awesome." In *Doctrine and Doxography: Studies on Heraclitus and Pythagoras*. Edited by Dirk Obbink and David Sider. Berlin and Boston: Walter de Gruyter, 2013, 3–32.

Schoenfeld, Alan. "When Good Teaching Leads to Bad Results: The Disasters of 'Well Taught' Mathematics Courses." *Educational Psychologist* 23, no. 2 (1988).

Trujillo, K. M., and O. D. Hadfield. "'Tracing the Roots of Mathematics Anxiety Through In-Depth Interviews with Preservice Elementary Teachers." *College Student Journal* 33 (1999): 219–32.

Turner, J. C., et al. "Working Memory, Comprehension, and Aging: A Review and a New View." *Psychology of Learning and Motivation* 22 (1988): 193–225.

Turner, William. "A Brief Introduction to Proofs." October 22, 2010, http://persweb.wabash.edu/facstaff/turnerw/Writing/proofs.pdf.

Young, Christina B., Sarah S. Wu, and Vinod Menon. "The Neurodevelopmental Basis of Math Anxiety." *Psychological Science* 23, no 5 (2012).

Young, D. A. B. "Ramanujan's Illness." *Notes and Records of the Royal Society London* 48 (1994): 107–19.

A NOTE ABOUT THE AUTHOR

Alec Wilkinson is the author of eleven books, including *The Ice Balloon*, *The Protest Singer*, *My Mentor*, *Mr. Apology*, and *The Happiest Man in the World*. Since 1980 he has been a contributor to *The New Yorker*. Before that he was a policeman in Wellfleet, Massachusetts, on Cape Cod, and before that he was a rock and roll musician. He lives in New York City with his wife and son.